Introduction to Workplace Safety and Health Management

A Systems Thinking Approach

Second Edition

AF372753

World Scientific Series On The Built Environment

Series Editor: Willie Chee Keong Tan *(National University of Singapore, Singapore)*

Published

Vol. 2   *Introduction to Workplace Safety and Health Management:*
          A Systems Thinking Approach
          Second Edition
          by Yang Miang Goh

Vol. 1   *Contract Administration and Procurement in the Singapore*
          Construction Industry
          Second Edition
          by Pin Lim

World Scientific Series on the Built Environment

Volume 2

Introduction to Workplace Safety and Health Management

A Systems Thinking Approach

Second Edition

Yang Miang Goh

National University of Singapore, Singapore

World Scientific

NEW JERSEY · LONDON · SINGAPORE · BEIJING · SHANGHAI · HONG KONG · TAIPEI · CHENNAI · TOKYO

Published by

World Scientific Publishing Co. Pte. Ltd.
5 Toh Tuck Link, Singapore 596224
USA office: 27 Warren Street, Suite 401-402, Hackensack, NJ 07601
UK office: 57 Shelton Street, Covent Garden, London WC2H 9HE

Library of Congress Control Number: 2020950194

British Library Cataloguing-in-Publication Data
A catalogue record for this book is available from the British Library.

World Scientific Series on the Built Environment — Vol. 2
INTRODUCTION TO WORKPLACE SAFETY AND HEALTH MANAGEMENT
A Systems Thinking Approach
Second Edition

Copyright © 2021 by World Scientific Publishing Co. Pte. Ltd.

All rights reserved. This book, or parts thereof, may not be reproduced in any form or by any means, electronic or mechanical, including photocopying, recording or any information storage and retrieval system now known or to be invented, without written permission from the publisher.

For photocopying of material in this volume, please pay a copying fee through the Copyright Clearance Center, Inc., 222 Rosewood Drive, Danvers, MA 01923, USA. In this case permission to photocopy is not required from the publisher.

ISBN 978-981-122-497-3 (hardcover)
ISBN 978-981-122-625-0 (paperback)
ISBN 978-981-122-498-0 (ebook for institutions)
ISBN 978-981-122-499-7 (ebook for individuals)

For any available supplementary material, please visit
https://www.worldscientific.com/worldscibooks/10.1142/11956#t=suppl

Desk Editor: Amanda Yun

Typeset by Stallion Press
Email: enquiries@stallionpress.com

About the Author

Goh Yang-Miang is an Associate Professor and Director at the Centre for Project and Facilities Management, Department of Building, School of Design and Environment, National University of Singapore (NUS). He holds both a B.Eng. and Ph.D. in Civil Engineering from NUS. Prior to his current appointment, Dr Goh was Deputy Head (Research) (2016–2019) and Deputy Programme Director (B.Sc. Project and Facilities Management) (2013–2016), Department of Building, NUS; Assistant Professor at the Department of Building, NUS (2012–2016); Senior Consultant at Det Norske Veritas, Singapore (2011–2012); Senior Lecturer at the School of Public Health, Curtin University, Western Australia (2007–2010); Engineer, Group Leader, Head, and Assistant Director (Investigations) at the Occupational Safety and Health Division at the Ministry of Manpower, Singapore (2005–2007); and Research Engineer/Research Assistant at NUS (2003–2004).

Prof Goh is also a Member of the External Review Panel on SAF Safety (ERPSS) (2017–present); Member of the Workplace Safety and Health Council Construction and Landscape Committee (2015–2017); Chairman/Co-Chairman of the Health and Safety Engineering Technical Committee, Institution of Engineers (IES), Singapore (2015–2018); and Academic Panel Member, IES Academy (2016–present). His current research interests are: Workplace safety and health; Systems thinking; Fall protection; Analytics, machine learning, computer vision and knowledge-based systems; and management of complex projects.

List of Abbreviations

Abbreviation	Full term
AARC	Approved Asbestos Removal Contractor
ACC	Approved Crane Contractors
ACoP	Approved Code of Practice
ATM	Automated Teller Machine
BBS	Behaviour-based Safety
DfS	Design for Safety
HSE	Health and Safety Executive
ISOM	Isomerisation
KPI	Key Performance Indicator
MOC	Management of Change
MOM	Ministry of Manpower
NEA	National Environment Agency
OHS	Occupational health and safety
PDCA	Plan-Do-Check-Act
PE	Professional Engineer
PEL	Permissible Exposure Levels
PM	Project Manager
PPE	Personal protective equipment
PTW	Permit-to-work
RA	Risk Assessment
RM	Risk Management
SHE	Safety, Health and Environment
SHS	Steel Hollow Section
STOP	Stop, Think, Observe and Plan
SWP	Safe Work Procedure

WSH	Workplace Safety and Health
WSHA	Workplace Safety and Health Act
WSHC	Workplace Safety and Health Council
WSHMS	Workplace Safety and Health Management System
WSHO	Workplace safety and health officer
WAH	Working-at-height

Contents

Chapter 5   Design for Safety   **139**

Chapter 10   Accident Case Studies   289

CHAPTER 1

Introduction

1.1 Importance of Workplace Safety and Health Management

Neglecting workplace safety and health[1] (WSH) management can cause accidents and ill health, which can lead to severe consequences for individuals, families, communities and organisations. However, as WSH incidents are uncertain events which may not occur even if work is conducted unsafely, many organisations do not see the importance of WSH. Managers may also neglect WSH management when there are other pressing issues related to revenue, schedule and client expectations that require their attention and resources.

This book is written for current and future managers overseeing and managing high risk workplaces like construction worksites, shipyards and factories. Even if a manager is not managing operations directly, s/he makes decisions that have significant influence on WSH. Thus, this book is also relevant to professionals like project managers working for clients, contracts managers, engineers and architects, who are involved in the design, selection and planning of products and operations. Knowledge of WSH management will also allow managers to identify contractors with a strong WSH competency. In addition, with legislations such as the WSH (Design for Safety) Regulations ('DfS Regulations'), all stakeholders are expected

---

[1] Also known as occupational safety and health (OSH) or occupational health and safety (OHS).

to be proactive in WSH management. Design for Safety (DfS) (also known as prevention through design, safe design and Construction (Design and Management)) promotes early consideration of safety and health hazards during the design phase of a construction project. Similarly, in other industries, considerations of safety during upstream stages have significant benefits. With early intervention, hazards can be more effectively eliminated or controlled, leading to safer workplaces and processes.

1.1.1 Accidents

Accidents happen, and when they happen, people suffer. However, many people simply do not register this simple fact. Even in relatively low risk environments like universities, there have been incidents of fire, which have caused property damage. For example, the National University of Singapore (NUS) had fires in August 2012, October 2012 and April 2014,[2] fortunately, these fires did not result in any major injury or fatality. In 2008, a tower crane collapsed in NUS,[3] killing three workers.

In Singapore, the Nicoll Highway collapsed on 20 April 2004 (Goh and Soon, 2014), and it remains one of the worst industrial accidents that Singapore has had. The Nicoll Highway collapse led to four deaths, and the body of the foreman, Mr Heng Yeow Peow, who saved several workers during the collapse, was never found. Though the four deaths were very significant, and the impact on the families of the victims was immeasurable, it must be remembered that the scale of the collapse of the Nicoll Highway, a busy road linking to the central area of Singapore,

---

[2] News articles on the fires can be seen online at https://sg.news.yahoo.com/fire-breaks-out- at-nus.html, https://sg.news.yahoo.com/small-fire-breaks-out-in-nus-lecture-theatre.html, and http://www.asiaone.com/singapore/fire-nus-engineering-faculty-building.

[3] Newsarticleavailableonhttp://www.asiaone.com/News/Education/Story/A1Story20080222-50958.html.

could have easily led to more fatalities. After the accident, senior managers of the main contractor, Nishimatsu (partner in the Nishimatsu-Lum Chang Joint Venture), were severely fined (Goh, 2008). The Professional Engineer (PE) cum project coordinator had made severe errors in the design of the diaphragm wall and strut system, which led to the collapse of the cut and cover tunnel. The PE from New Zealand was fined S$160,000 in April 2006 and banned from practice for two years. In addition, Nishimatsu was fined S$200,000, the design manager was fined S$160,000, and the project director was fined S$120,000. The former Land Transport Authority (LTA) project director was fined S$8,000 for failing to exercise due diligence in monitoring excavation works and assessing readings of soil monitoring instruments. The Nicoll Highway station project was delayed by four years and millions of dollars were spent on reconstructing the highway and adjacent roads. The Nicoll Highway collapse showed not only that accidents cause human suffering and project failure, but that managers, project managers and executives can be taken to task for not ensuring safety.

In recent years, the Singapore Ministry of Manpower (MOM) has been taking project managers and appointment holders to task for their failure to ensure safety on site. In one instance, a project coordinator cum lifting supervisor was sentenced to four weeks' imprisonment. Thus, managers must be competent in and committed to WSH management.

1.1.2 Ill health

Unlike accidents, ill health takes years to develop and the illness may not be clearly linked to the work that the victim conducted many years ago. According to the International Labor Organization (2009), each year, about 2.3 million people die from work-related accidents and diseases. Out of the 2.3 million deaths, about 360,000 are due to accidents, while the remaining 1.95 million are fatal work-related diseases. Even though most of the deaths occur in developing countries and the statistics include all types of workplaces, the astonishing numbers signal the impact of occupational diseases. Some of the occupational health issues common across industries include noise induced deafness, skin conditions like dermatitis, musculoskeletal disorders, cancer, and occupational asthma.

1.1.3 Environmental pollution

Environmental pollution includes local pollution problems, like noise and odour, and global pollution issues, like the depletion of natural resources, the inefficient use of energy, carbon dioxide and greenhouse gas emissions, air pollution, water pollution, land pollution, and the loss of biodiversity. Both local and global environmental pollution are undesirable, but global pollution, particularly with regard to climate change-related issues, is becoming more and more important. With the global average temperature on the rise and more extreme and erratic weather occurring, the industry has to study the environmental aspects (synonymous to hazards in WSH terminology) and impacts (synonymous to accidents or ill health) carefully.

Environmental problems are closely related to WSH. Accidents such as major fires and structural collapses can release substances and dust into the environment and will also lead to pollution and other environmental impacts. As in the case for WSH, the Singapore National Environment Agency (NEA) has been very strict in their enforcement of Environmental Protection and Management legislations, and workplaces have to ensure they have a systematic approach for environmental management.

1.1.4 Cost of incidents

WSH incidents are costly. According to Bird *et al.* (2003) (see Figure 1.1), for each dollar of insured cost, there is another

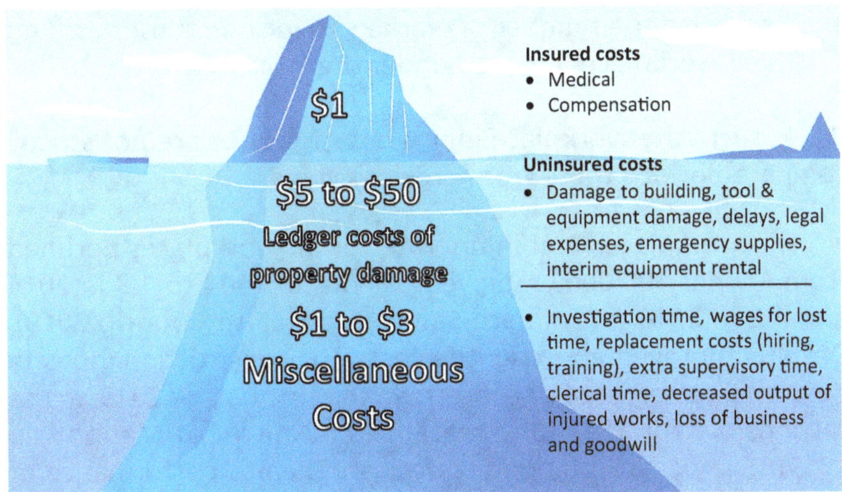

Figure 1.1   The cost of incidents

US\$6 to US\$53 of uninsured costs. Just like a ship captain who only sees the tip of the iceberg, many organisations are focused on the insured costs, but they may not have noticed the massive hidden uninsured costs. The actual ratio between insured and uninsured cost will vary from incident to incident, but the key message is that incidents are costly, and it is worthwhile to invest in incident prevention.

1.2 Workplace Safety and Health Statistics

WSH statistics help us understand how different industries are performing in terms of WSH. The WSH statistics in Singapore are reported by the Workplace Safety and Health Institute (WSHI). Before interpreting the statistics, it is important to know that as of 2014, the WSH statistics report:

a. Included work-related traffic injuries
b. Reclassified work-related back injuries due to ergonomic risks, such as work-related musculoskeletal disorders, as occupational diseases, and

c.   Expanded the number of workers to include those working in all workplaces covered under the WSH Act.

As such, the workplace injury rates for 2014 are not strictly comparable with those from previous years.

The workplace fatal injury rate for all workplaces declined from 4.9 per 100,000 employed persons in 2004 to 1.8 in 2014 and 1.2 in 2018 (Workplace Safety and Health Council, 2019). Despite improvements over the years, the construction industry has remained an area of concern. In the first half of 2019, it was responsible for 35% (6) of fatalities (Ministry of Manpower, 2019). The persistently poor safety and health performance of the construction industry made it important for all stakeholders to do their part to improve the situation.

Nevertheless, other industries, such as the process and energy industries, cannot be complacent; because unlike the construction industry or marine industry, the process industry can cause major accidents with extremely severe consequences, in the range of hundreds or thousands of injuries and deaths. For example, the Bhopal gas leak disaster in India on 3 December 1984, killed at least 4,000 people. In addition, thousands of others were injured, and many more were plagued with ill health. Major accidents like the Bhopal gas leak are stark reminders that major hazard industries must have high WSH standards.

1.3 Workplace Safety and Health Act

The Singapore Workplace Safety and Health Act (WSHA) is heavily influenced by the UK Health and Safety at Work, etc. Act 1974, and similar legislations in Europe and Australia. As an illustration of the performance-based approach to WSH legislation, the WSHA will be discussed herein.

The WSHA was enacted in 2006 after three major accidents in 2004: the Nicoll Highway collapse on 20 April 2004, the collapse of steel latticework at the Fusionopolis Building worksite on 29 April 2004, and the fire on *Almudaina* at Keppel Shipyard on 29 May 2004. The overall intent of the WSHA was to improve the safety culture of the industry and encourage stakeholders to take reasonable practicable steps to improve WSH proactively. The WSHA is based on three basic principles:

1. Reducing risks at the source by requiring all stakeholders to eliminate or minimise the risks they create;
2. Instilling greater ownership of safety and health outcomes within the industries; and
3. Preventing accidents through higher penalties for compromises in safety management.

In contrast to the repealed Factories Act (a prescriptive legislation that details WSH requirements specifically), the WSH Act is a performance-based legislation which requires the industry to conduct risk assessment (RA) so as to proactively identify the hazards, evaluate the hazards, determine controls for the hazards, and implement and review the hazards and controls. One key feature of a performance-based approach is the concept of "as low as reasonably practicable" (ALARP). Based on the case of Public Prosecutor v Hong Jun Development Pte Ltd [2017] SGMC 68, 'reasonably practicable' contains the following principles:

1. The risk of accident has to be weighed against the measures necessary to eliminate the risk, including the cost involved;
2. "The term 'reasonably practicable' means that stakeholders need only take preventive measures which are proportionate to the potential impact of the hazards at the workplace."

3.   "...if a precaution is practicable it must be taken unless in the whole circumstances that would be unreasonable. And as men's lives may be at stake, it should not lightly be held that to take a practicable precaution is unreasonable."

In essence, "reasonably practicable" means that the benefits in reduction of risk must be weighed against the costs (e.g. time, money and resources) of implementing the controls. The weighing or evaluation of costs versus benefits should make reference to regulations, approved codes of practice (ACoP), industry standards, guidelines and norms to define what is reasonable and practicable.

Compared to the prescriptive approach in the repealed Factories Act, the performance-based regime is more sustainable because it is not possible for the government to keep creating new legislation to cover different hazards and controls. The emphasis on WSH management is an important one because penalising a company only when it has accidents or ill health is reactive and workers would already be infected, injured or killed. Promoting effective WSH management reduces the risk of incidents and pollution. The duty holders covered in the WSH Act and their key duties are highlighted in Table 1.1.

Each duty holder has a role to play in preventing accidents and ill-health and can be taken to task for failing to perform their duties. A set of subsidiary legislations are written under the WSH Act, and to support the subsidiary legislations and the WSH Act, a wide range of standards were developed. Standards approved by the MOM are known as ACoP, and they have higher standing in the courts compared to non-approved codes of practice. The list of ACoPs is published in the government Gazette.

The WSH Act imposes a maximum penalty of S$500,000 for a corporate body with a first conviction. For a repeat offender,

Table 1.1   Duties of duty holders under WSHA (adapted from Ministry of Manpower, 2016b)

Duty Holder	Definition	Key Duties
Employer	Section 6(1) — "employer" means a person who, in the course of the person's trade, business, profession or undertaking, employs any person to do any work under a contract of service.	Section 12 — An employer must protect the safety and health of his employees or workers working under his direction, as well as persons who may be affected by their work. The employer must: • Conduct risk assessments to identify hazards and implement effective risk control measures. • Make sure the work environment is safe. • Make sure adequate safety measures are taken for any machinery, equipment, plant, article or process used at the workplace. • Develop and implement systems for dealing with emergencies. • Ensure workers are provided with sufficient instruction, training and supervision so that they can work safely.

(Continued)

Table 1.1   (*Continued*)

Duty Holder	Definition	Key Duties
Principal	Section 4(1) — "principal" means a person who, in connection with any trade, business, profession or undertaking carried on by him, engages any other person otherwise than under a contract of service — (a) to supply any labour for gain or reward; or (b) to do any work for gain or reward.	Section 14 — A principal must ensure that the contractor he engaged: • Is able to perform the work they are engaged for. • Has made sure that any machinery, equipment, plant, article or process that is used at work is safe. However, if the principal instructs the contractor or the workers on how the work is to be carried out, his duties will include the duties of an employer.
Occupier	Section 4(1) — "occupier", in relation to any premises or part of any premises, means: (a) in the case of a factory where a certificate of registration has to be obtained in relation to the premises pursuant to any regulations — the person who is, or is required to be, the holder of the certificate;	Section 11 — An occupier must ensure that the following are safe: • The workplace. • All pathways to and from the place of work. • Machinery, equipment, plants, articles and substances. The occupier must ensure that the above does not pose a risk to anyone within his premises, even if the person is not his employee.

	(b) in the case of a factory where a notification has to be submitted in relation to the factory pursuant to any regulations — the person who is named in the notification, or is required to submit a notification; and (c) in the case of any other premises — the person who has charge, management or control of those premises either on his own account or as an agent of another person, whether or not he is also the owner of those premises;	Section 19 — The occupier may also be responsible for the common areas used by his employees and contractors. Common areas include the following: • Electric generators and motors. • Hoists and lifts, lifting gears, lifting appliances and lifting machines. • Entrances and exits. • Machinery and plants.
Manufacturer or supplier of machinery and equipment or hazardous substances	Not defined in WSHA	Section 16 — A manufacturer or supplier must ensure that any machinery and equipment or hazardous substances he provides are safe. He must: • Provide information on health hazards and how to safely use the machinery, equipment or hazardous substance. • Examine and test the machinery, equipment or hazardous substance to ensure that it is safe for use. • Provide results of any examinations or tests of the machinery, equipment or hazardous substances.

(Continued)

Table 1.1   (*Continued*)

Duty Holder	Definition	Key Duties
Installer or erector of machinery	Not defined in WSHA	Section 17 — The installer or erector of machinery must ensure that the machinery and equipment that he has erected, installed or modified is safe and without safety or health risks when properly used.
Employee	Section 6(1) — "employee" means any person employed by an employer to do any work under a contract *of* service. (Note that "contract *for* service" refers to "an independent contractor, such as a self-employed person or vendor, is engaged for a fee to carry out an assignment or project." (Ministry of Manpower, 2017)	Section 15 — An employee must: • Follow the workplace safety and health system, safe work procedures or safety rules implemented at the workplace. • Not engage in any unsafe or negligent act that may endanger himself or others working around him. • Use the personal protective equipment provided to him to ensure his safety while working. He must not tamper with or misuse the equipment.
Self-employed	Not defined in WSHA	Section 13 — A self-employed person is required to take measures to ensure the safety and health of anyone in the workplace who may be affected by his work.

this is increased to S$1 million (Ministry of Manpower, 2016a). For individuals, the maximum penalty is S$200,000 for a first conviction and S$400,000 for a repeat offender. Individuals can also be imprisoned for a maximum of two years. The MOM can also impose composition fines instead of prosecuting. Each offence may be compounded to a sum not more than half the maximum fine prescribed for the offence or S$5,000, whichever is lower.

1.4 Overview of Workplace Safety and Health Management

Cost, quality, competition, profit, timeline, environmental pollution and WSH are different types of challenges that managers need to handle on a daily basis. On the surface, these challenges are unrelated and independent. However, the reality is that they are intertwined and inter-dependent (see Figure 1.2).

For example, in order to push for project progress, a project manager may instruct workers to bypass safety and health procedures. A safety inspection may be cancelled, or the safety barricades may not be installed to save time. This leads to higher risk, and if accidents happen, the workplace will be

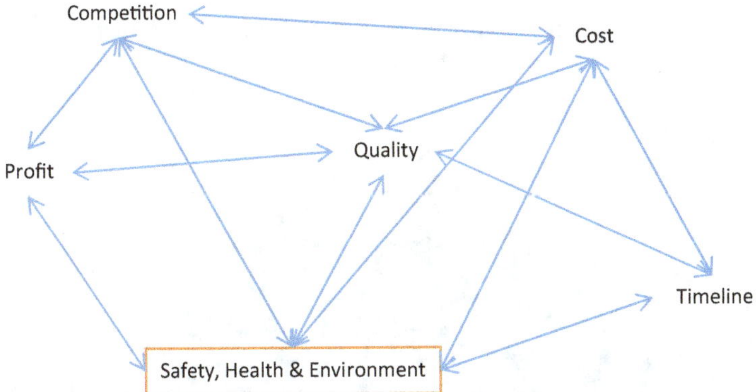

Figure 1.2   The inter-dependent nature of all management issues

subjected to investigations, leading to delays and loss of profit. Quality and WSH are related because when workers rush their work, not only are they more likely to make safety-related errors, they can easily make mistakes leading to quality issues. Poor workmanship can also mean future rework, and that means more time lost. However, WSH incidents do not occur every time WSH procedures and measures are breached. This is because there is always an element of uncertainty or luck. Thus, managers tend to focus on certain — typically, more urgent — issues that they are facing. This frequently leads to the neglect of WSH management. To counteract this tendency, it is important for managers to focus on the fundamentals of WSH management, which are aligned with the principles of good management and planning.

The three basic principles of the WSH Act show the importance of WSH management — in particular, Principle 3, which implies that companies can be penalised even if there is no accident on site. Organisations are expected to manage safety proactively and all stakeholders have to be involved. The key principles of the WSH Act are also aligned with the well- known accident pyramid or ratio (see Figure 1.3), which was developed based on a set of accident data that F. E. Bird collected (Bird *et al.*, 2013). The pyramid shows that for each major injury there are disproportionately more minor injuries, property losses and near hits. This implies that if management continuously learns

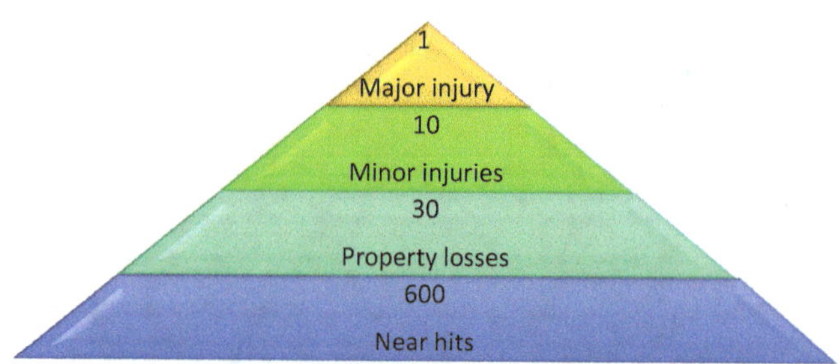

Figure 1.3   F. E. Bird's accident pyramid

from near hits, property losses and minor injuries, it is possible to prevent major injuries from happening. Nevertheless, after the BP Texas oil refinery explosion in 2015, many academics warned that the concept of an accident pyramid can also be misleading. Major accidents, e.g. the major explosion of a plant and the collapse of a large structure, can have very different causes from minor injuries, property losses and near hits, especially if the nature of the incident is very individual, e.g. a worker cut his finger or fell when walking. The accident pyramid should only be taken as a guide, but its emphasis on proactively identifying incidents of low consequence to learn from is still important, as long as managers are conscious of the difference between major accidents and minor individual accidents. In addition, the ratio of 1:10:30:600 is not fixed and can change when applied to different data sets.

WSH management is based on the Plan–Do–Check–Act (PDCA) cycle (Figure 1.4), which is also adopted in quality management, environmental management and risk management. During *Plan*, organisations must proactively identify the possible hazards (WSH) or aspects (environmental) that can lead to accidents and ill health or environmental impacts. A series of controls or programmes will be identified during planning. During *Do*, the organisation will implement the controls and programmes to control the risk posed by the hazards and aspects. The effectiveness of the planning and doing processes

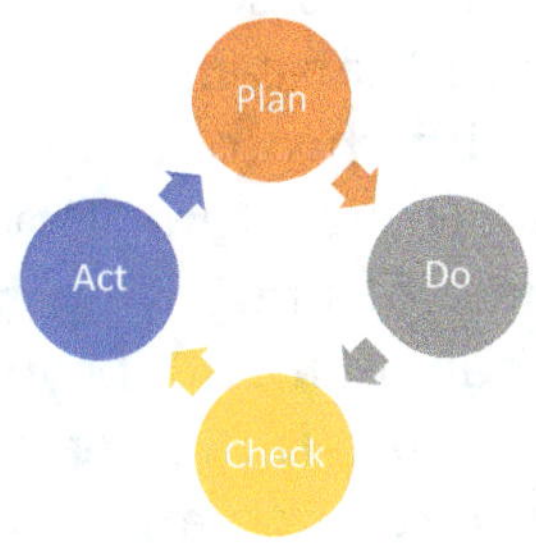

Figure 1.4   The Plan–Do–Check–Act (PDCA) cycle

will be *Checked.* If there are gaps, the organisation will *Act* on those identified issues for continuous improvement. The cycle continues to ensure progressive improvement.

In the context of WSH management, there are several standards and guidelines available to help organisations understand the policies, structure, procedures and actions needed to maintain a management system that continually improves WSH performance. Many of these WSH management system standards are structured based on the PDCA cycle. The BS OHSAS 18001 Occupational Health and Safety Management Systems Standard (British Standards Institute, 2007) was one of the most commonly adopted WSH management standards internationally. The OHSAS 18001 is similar to ISO 14001 (British Standards Institution, 2015), but the latter is designed for environmental management. The ISO 45001:2018 (ISO, 2018) replaced BS OHSAS 18001 in 2018. ISO 45001 will be discussed in Chapter 6.

1.5 Workplace Safety and Health Systems Thinking

Systems thinking (or system dynamics) is a discipline to help managers understand inter-relationships between system components so as to better manage complex systems, which can behave unpredictably. One of the fundamental principles of systems thinking is the importance of looking beyond the direct and immediate causes of events. A systems thinker will seek to understand the patterns of behaviour of the system, system components and stakeholders so as to uncover the underlying factors and structure that caused the event to occur. This is very much aligned with the concept of root cause analysis commonly found in WSH literature. However, systems thinking takes a broader view of underlying factors and structures than WSH root cause analysis. It goes beyond the WSH management system and safety culture, to see the organisation as a whole system. Furthermore, systems thinking seeks to understand unpredictable system behaviour arising from interactions between system components. When necessary, the boundary of the system being evaluated can also be expanded to cover areas beyond the organisation so as to facilitate the identification of underlying factors and leverage points.

This holistic view of WSH management is essential in improving WSH performance, because WSH issues may have non-WSH origins. In addition, systems thinking emphasises that complexities arise from simple dynamics such as the circular dependence between system parameters and components, and delays (Goh *et al.*, 2010). We must note, however, that system dynamics literature tend to overemphasise quantitative approaches to model these complexities through stocks and flow models. In contrast, this book adopts the key principles of systems thinking and takes a practical and qualitative approach to their application in WSH management.

Review Questions

1. Why is Safety, Health and Environmental (SHE) management important?
2. In contrast to workplace accidents, why are occupational health and environmental pollution more frequently neglected?
3. Why is it important for all stakeholders to emphasise the importance of WSH management?
4. What are the implications of F. E. Bird's accident pyramid?
5. Describe the PDCA cycle in the context of WSH management.
6. What is systems thinking and how is it applicable to WSH management?

References

Bird, F. E., Germain, G. L., and Clark, M. D. (2003). *Practical loss control*, Det Norske Veritas (U.S.A.), Inc., Duluth, Georgia.
British Standards Institute (2007). "BS OHSAS 18001:2007 Occupational health and safety management systems — Requirements." BSI, London.

British Standards Institution (2015). "BS EN ISO 14001:2015 Environmental management systems — Requirements with guidance for use." British Standards Institution, London.

Goh, C. L. (2008). "NZ engineer suspended from practice for two years." *Straits Times*, Singapore Press Holdings Limited, Singapore.

Goh, Y. M., Brown, H., and Spickett, J. (2010). "Applying systems thinking concepts in the analysis of major incidents and safety culture." *Safety Science*, 48, 302–309.

Goh, Y. M., and Soon, W. T. (2014). *Safety Management Lessons from Major Accident Inquiries*, Pearson, Singapore.

International Labor Organization (2009). "Facts on safety and health at work." International Labor Organization, Geneva.

ISO. (2018). *ISO 45001:2018 Occupational health and safety management systems.*

Ministry of Manpower (2016a). "General penalties." http://www. mom.gov.sg/ workplace-safety-and-health/workplace-safety-and-health-act/liabilities-and- penalties. (Nov 16, 2017).

Ministry of Manpower (2016b). "WSH Act: responsibilities of stakeholders."

http://www.mom.gov.sg/workplace-safety-and-health/workplace-safety- and-health-act/responsibilities-of-stakeholders. (Nov 17, 2017).

Ministry of Manpower (2017). "Contract of service." http://www.mom. gov.sg/employment-practices/contract-of-service. (Nov 14, 2017).

Ministry of Manpower (2019). "Workplace safety and health reports and statistics." https://www.mom.gov.sg/workplace-safety-and-health/wsh-reports-and-statistics. (8 January 2020).

Workplace Safety and Health Council (2019). "WSH2028 — Singapore Aims To Be One of The Safest Workplaces In The World By 2028" https://www.wshc.sg/wps/portal/!ut/p/a1/jY89D4IwEIZ_ iwMrd3yIxq1xkCjGAVToYsBgwSBt2kr_vsjkIOht7-V5cvcChRRom-3c1y3XN27x5ZxpcooPvuyTG3SZMHCTuHsNk4TIx4PdANg6sT95_ Po4MwV_-GegUMnwwABMntkBZw4uhbkbawlsyoLK8IbKU9IP 260proVYWWmiMsY2qrrZifRDKQsGI_q5WXGIIPwwQj2OK93n- TRWT2AgTnMHU!/dl5/d5/L2dBISEvZ0FBIS9nQSEh/?action=cmsPu blicView&cmsId=C-2019040511721&tabId=C-2019040511722. (25 August 2020).

CHAPTER 2

Incident Causation

2.1 Introduction

Prior to the discussion on incident causation models, it is necessary to first establish the difference between accidents, incidents and events (see Figure 2.1). An "event" is defined as an episode of something that happened, which may or may not have any consequences. Examples of an event include an overseas trip, a meeting and watching a movie. "Accidents" are defined as unexpected and undesirable events that result in negative consequences such as injuries, fatalities, illnesses, property damage and financial loss. Accidents are a subset of incidents, which include near misses or, more accurately, near hits, which are undesirable events that could have resulted in negative consequences if the negative consequences had not been averted by chance or successful safety or emergency measures. Based on the above definitions, "incidents" will include workplace accidents, environmental pollution and occupational diseases.

When an event occurs, different people may view the same event differently and can derive different insights from the same event. The differences in insights can then lead to different actions with varying levels of effectiveness. The differences in perception and insights are influenced by differences in mental models or paradigms, which are our internal representations of how the world works. This internal representation is made up of simplified concepts about the real world and the relationships between the concepts. Mental models are like maps, which are simplifications of the real world, but they are useful in guid-

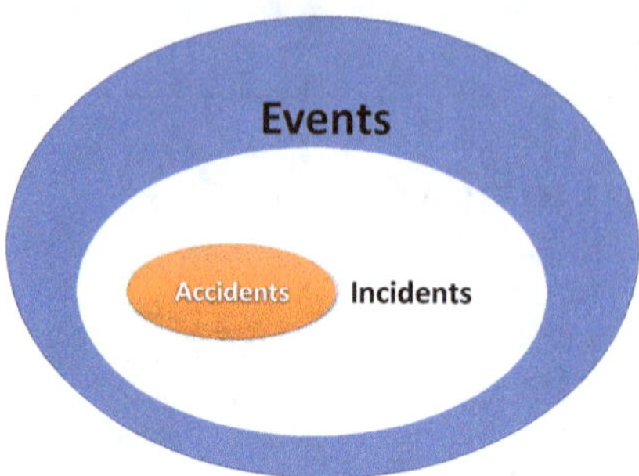

Figure 2.1   Defining accidents, incidents and events

ing our actions. Thus, when we manage workplace safety and health (WSH), it is important to have a suitable incident causation model that guides us in the way that we prevent and investigate accidents. A good incident causation model will help us derive suitable insights and actions to prevent incidents.

There have been many incident causation models developed over the years. This chapter will present the Domino Theory and Energy Transfer Model as traditional incident causation models, introduce the Loss Causation Model (LCM) and Swiss Cheese Model (SCM) as systemic incident causation models, and discuss the importance of systems thinking in incident causation.

2.2 Domino Theory

The Domino Theory by Heinrich (1936) (see Figure 2.2) is one of the earliest incident causation models, but it is now considered unsuitable. It focuses on how the unsafe acts, carelessness and anti-social behaviours (e.g., alcoholism) of individuals contribute to accidents. According to the Domino Theory, the fundamental cause of an accident, as represented by the last block of the domino, is the "social environment and inherited behaviour" of the victim and/ or workers responsible for the accident. The Domino Theory reflects

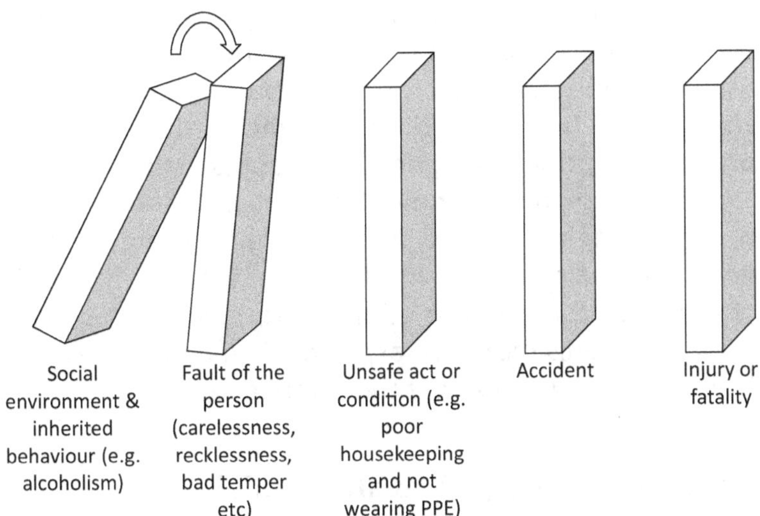

Figure 2.2   An adapted version of Heinrich's Domino Theory; "PPE" in the figure refers to Personal Protective Equipment

the prevalent mental models of the past, which were usually focused on the equipment, environmental factors and human error (including violations). Such mental models still persist today. For example, in July 2017, after the collapse of a viaduct that was under construction in Singapore, the media (Christopher Tan, 2017) highlighted human error as the likely cause of the accident without any discussion on the more fundamental management-related issues. Blaming fault of person, social environment and inherited behaviour (non-management related factors) as the fundamental cause of incidents can lead to management blaming the workers for incidents and ending investigations before more fundamental or underlying factors are uncovered. The failure to remove or mitigate the underlying factors can then lead to the recurrence of incidents. However, it must be noted that the individuals involved are also accountable for their own actions and that it is important for workers to cooperate with their employers and comply with WSH rules and regulations.

The Domino Theory is now considered unsuitable because organisations have come to realise that there are other more fundamental underlying factors contributing to human error, unsafe acts and unsafe conditions. Some of these underlying factors include poorly designed procedures, production pressure, error-provoking equipment design and lack of management commitment to WSH. The general agreement among modern incident causation models is that, even though incidents are caused by unsafe equipment, environment and/or behaviours, organisations are expected to manage these direct causes in a proactive and *systemic* manner.

2.3 Haddon's Energy Transfer Model

The Energy Transfer Model (ETM) (Haddon Jr, 1973) provides a practical understanding of incidents and control measures. The ETM (see Figure 2.3) traces the interactions between an energy source or hazardous substance and a person or property and

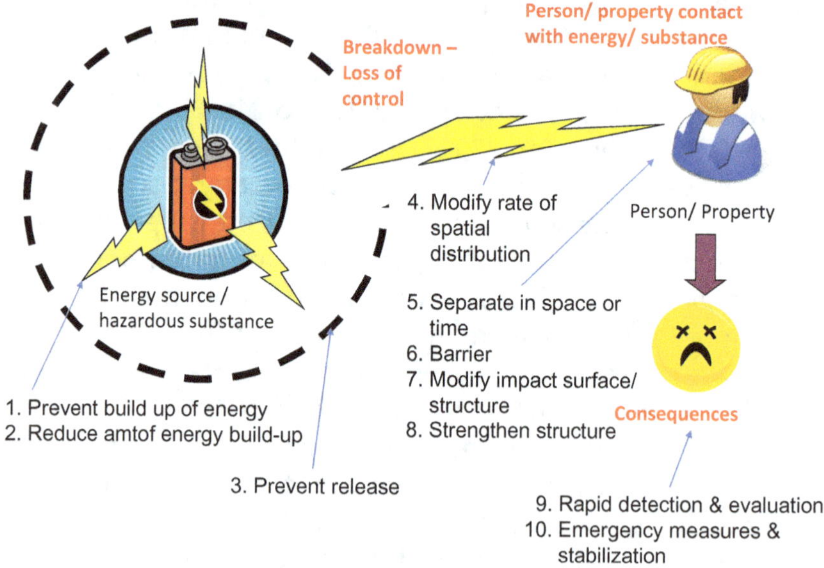

Figure 2.3   Adapted from Haddon's Energy Transfer Model (1973)

the resulting consequences. Ten counter-measures or controls are identified based on the model. The ten incident controls identified by Haddon Jr. (1973) are:

1. To prevent the initial marshalling of the form of energy
2. To reduce the amount of energy marshalled
3. To prevent the release of the energy
4. To modify the rate of spatial distribution of release of energy from its source
5. To separate in space or time the energy being released from the susceptible structure
6. To separate the energy being released from the susceptible structure by interposition of a material barrier
7. To modify the contact surface, subsurface, or basic structure which can be impacted
8. To strengthen the structure which might be damaged by the energy transfer
9. To move rapidly in detection and evaluation of damage and to counter its continuation and extension
10. All those measures which fall between the emergency period following the damaging energy exchange and the final stabilisation of the process

The abbreviated forms of the ten controls are inserted into Figure 2.3. The ETM is useful in risk assessment (RA) and incident investigation. In fact, many of the incident causation models incorporate some of the concepts highlighted in the ETM. However, ETM is not a systemic model, which may lead to lack of emphasis on more fundamental or *systemic* causes of WSH incidents.

2.4 Systemic Models

Incident causation models that highlight the role of the organisation, its policies, procedures, structure and culture in the causation of an incident are classified as *systemic models*. In fact, there are also models that go beyond organisations to consider the influence of the industry, government and even beyond, but this book is focused on the prevention of incidents at the organisational level, and the influence of factors beyond the organisation are not discussed in detail.

Numerous systemic models have been developed over the years, for example, the Management Oversight Risk Tree (MORT) (Johnson, 1980), the contributing factors in accident causation (CFAC) model (Sanders and Shaw, 1988), the SCM (Reason, 1997), the LCM (Bird *et al.*, 2003), and the Modified Loss Causation Model (MLCM) (Chua and Goh, 2004). The Event Causation Technique (ECT), which is an extension of the MLCM, is the main systemic incident causation model that will be discussed, and it will be presented in detail in the following chapter. These models implicitly or explicitly reinforce the concept of multiple-causation, where the cause of an incident does not lie in a single line of causation, but often branches into various chains of factors.

Figure 2.4 captures the gist of systemic models, which indicates that the basic characteristics of a system (underlying factors) are fundamental to the occurrence of events. Events, as explained earlier, can be any episode of "things happening", including incidents. "Patterns" refer to patterns of events over time or patterns across entities like people, machines or organisations.

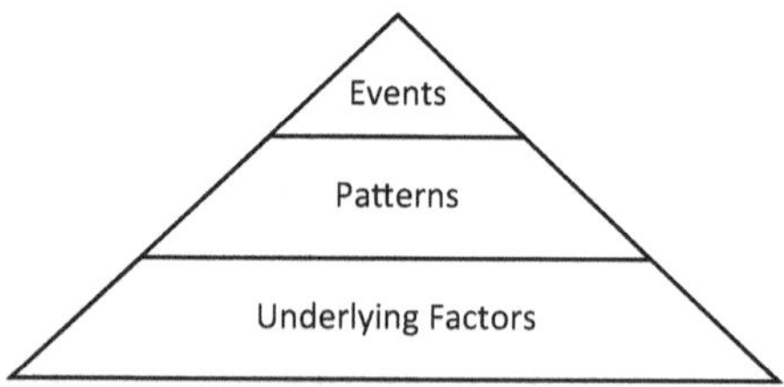

Figure 2.4   Systemic view of events; adapted from Senge (2006)

Finally, "underlying factors" refer to management system (e.g. ISO 45001 and ISO 14001) weaknesses, leadership inadequacies or organisational culture factors. The pyramid in Figure 2.4 indicates that events and incidents do not "just happen"; they arise due to certain systematic patterns of the way things are done in the organisation or certain characteristics of the organisation. Some examples of patterns can be the declining number of and competency of engineers across time, and the pattern of unsafe behaviour (e.g. not putting on their harness and lanyards) across workers. These observable patterns are usually related to poor management and organisational culture. Poor management refers to a weak management system, which does not have an effective Plan–Do–Check–Act (PDCA) cycle. For example, an organisation that does not have a systematic procedure for conducting RA, and this was not identified during periodic management system audits. The lack of RA then led to hazards not being assessed and no control being implemented to reduce the risk of the hazards. This results in high risk hazards, like open edges, being frequently observed on site. Which may finally lead to a worker falling from height (an event) when working near one of the open edges.

2.5 Loss Causation Model

The Loss Causation Model (LCM) (Bird *et al.*, 2003) has a similar structure to the Domino Theory, but it represents an important shift in mental models. With reference to Figure 2.5, the key difference between the Domino Theory and the LCM is that "lack of management control" is deemed to be the most fundamental cause of incidents and losses in the LCM. Coming from the left of the model, lack of management control (inadequate management system, inadequate standard or inadequate compliance to the system and standard) can lead to basic causes of incidents. As in the case of the Domino Theory, basic causes or root causes refer to personal factors and job or system factors. The former is defined as physiological and psychological factors,

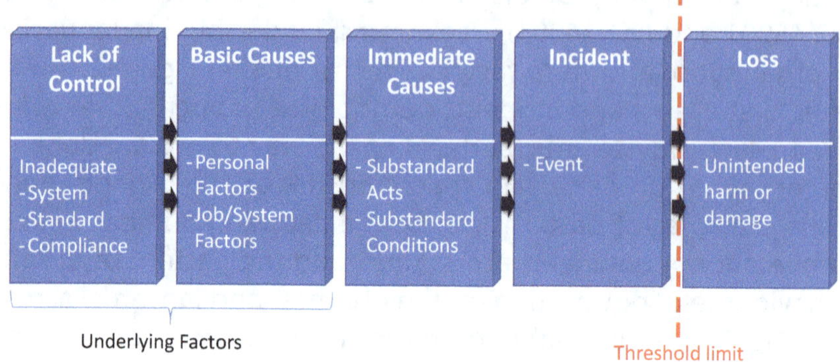

Figure 2.5   Loss Causation Model; adapted from Bird *et al.* (2003)

while job or system factors refer to factors such as leadership, work standards, RA and maintenance.

Incidents are directly caused by immediate causes, which include substandard acts and substandard conditions. "Acts" refer to observable human behaviours, and "conditions" refer to physical conditions such as the condition of equipment, material, structure and environment. An act or condition is considered substandard when it fails to meet a stipulated standard set by the organisation. This implies that organisations are expected to set comprehensive standards to cover behaviours and workplace conditions. Immediate causes are generally observable and they are more easily determined compared to basic causes.

"Loss" has a broader definition than human injuries and fatalities. Losses include property loss, reputation loss and environmental impact. Following Haddon's ETM, losses arise from incidents when a certain protective threshold has been exceeded. If a person is struck by a moving object, for example, a forklift travelling at extremely slow speed, the collision may not exceed the person's capacity to absorb the kinetic energy exerted by the forklift. In such a situation, the incident would remain as a near-hit because the injury threshold of the person has not been exceeded and there would not be any loss.

To better illustrate the LCM, Table 2.1 captures the possible classifications under each of the main blocks in the LCM. It must be noted that these classifications are not comprehensive, and are written in a generic fashion. They are useful prompters to help people consider possible causes and factors during incident investigation and classification, but incident specific information

Table 2.1   Loss causation model; adapted from Bird *et al.* (2003)

Losses from Incidents		
• Injured worker time	• Co-worker time	• Leader time
• General losses (lost production time, decreased effectiveness of employees and loss of business goodwill, etc.)	• Property losses	• Other losses
Event/Incident Types		
• Struck against (running or bumping into)	• Struck by (hit by moving object)	• Fall to lower level (either the body falls or the object falls and hits the body)
• Fall on same level (slip and fall, trip over)	• Caught in (pinch and nip points)	• Caught on (snagged, hung)
• Overstress or overexertion or overload	• Caught between (crushed or amputated)	• Contact with (any harmful energy or substance)

(Continued)

Table 2.1 (*Continued*)

• Release of (any harmful energy or substance)		
Immediate Causes		
Substandard Acts or Practices		
• Operating equipment without authority	• Failure to warn	• Failure to secure
• Operating at improper speed	• Making safety devices inoperable	• Using defective equipment
• Using equipment improperly	• Failing to use personal protective equipment properly	• Improper loading
• Improper placement	• Improper lifting	• Improper position for task
• Servicing equipment in operation	• Horseplay	• Under influence of alcohol and/or other drugs
• Failure to follow procedure or policy or practice	• Failure to identify hazard or risk	• Failure to check or monitor
• Failure to react/correct	• Failure to communicate or coordinate	
Substandard Conditions		
• Inadequate guards or barriers	• Inadequate or improper protective equipment	• Defective tools, equipment or materials
• Congestion or restricted action	• Inadequate warning systems	• Fire or explosion hazards

Table 2.1   (*Continued*)

• Poor housekeeping; disorder	• Hazardous environmental conditions: gases, dusts, smokes, fumes, vapours	• Noise exposure
• Radiation exposure	• Temperature extremes	• Inadequate or excess illumination
• Inadequate ventilation	• Inadequate instructions or procedures	
Basic Causes		
Personal Factors		
• Inadequate physical or physiological capability	• Inadequate mental/ psychological capability	• Physical or physiological stress
• Mental or psychological stress	• Lack of knowledge	• Lack of skill
• Improper motivation		
Job/System Factors		
• Inadequate leadership and/or supervision	• Inadequate engineering	• Inadequate purchasing
• Inadequate maintenance	• Inadequate tools and equipment	• Inadequate work standards
• Wear and tear	• Abuse or misuse	

(Continued)

Table 2.1   (*Continued*)

Lack of Control (System Elements)		
• Leadership and Administration	• Leadership training	• Planned inspections and maintenance
• Critical task analysis	• Incident investigation	• Performance observation
• Emergency preparedness	• Rules and work permits	• Incident analysis
• Knowledge and skill training	• Personal protective equipment	• Health and hygiene control
• System evaluation	• Engineering and change management	• Personal communications
• Team communications	• General promotion	• Hiring and placement
• Materials and services management	• Off-the-job safety	

must accompany the classification to provide the background information needed for someone to understand the reason for the classification.

2.6 Swiss Cheese Model

The Swiss Cheese Model (SCM) (Reason, 1997) shown in Figure 2.6 is used in many industries, such as the oil and gas, aviation, mining and nuclear industries. Based on the model, consequences are the result of several conditions: (1) the presence of a threat or danger that emits an accident trajectory, represented by the laser beam; (2) the presence of holes in the barriers created by active failures and/or latent conditions; and

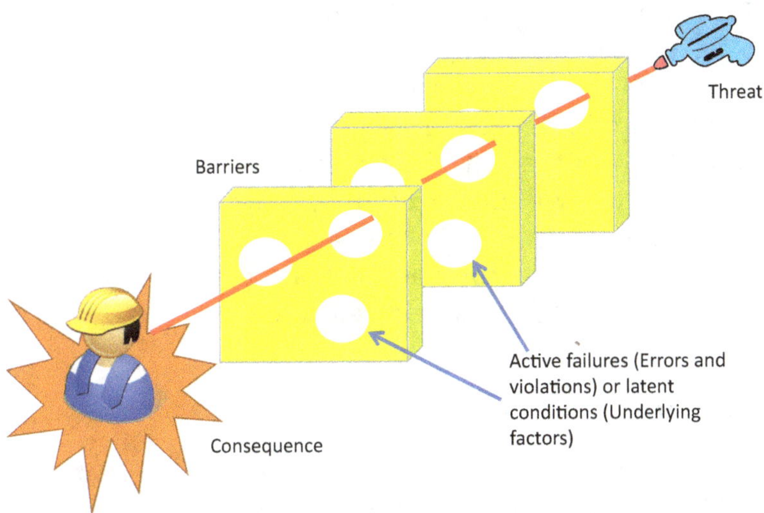

Figure 2.6   Swiss Cheese Model; adapted from Reason (1997)

(3) the alignment of the holes to allow the accident trajectory to hit the 'target', such as a property, equipment or person. The barriers are synonymous with the controls in the LCM and active failures are essentially the immediate causes. Latent conditions are like basic causes in the LCM. According to Reason (1997), active failures are the errors and violations committed at the 'sharp end' of the system; that means by frontline workers, pilots, maintenance personnel and operators. Active failures have relatively immediate and short-lived effects. Latent conditions are like pathogens to the human body, e.g. poor design, gaps in supervision, undetected manufacturing defects or maintenance failures, unworkable procedures, clumsy automation, shortfalls in training, inadequate tools and equipment. These latent conditions can be present in the organisations many years before they interact with active failures and local situations to defeat the different layers of barriers in the organisation. These latent conditions originate from high level decisions of managers, designers, manufacturers, regulators and governments.

Even though there are differences between the LCM and the SCM, both models emphasise that incidents and losses will occur if there is a lack of management control. However, the SCM highlighted safety culture as an important aspect of latent conditions, which the LCM did not cover explicitly. The concept of safety culture will be covered in Chapter 7. The SCM also emphasised that the holes in the barriers are inevitable, so the key is to reduce these holes as much as possible and to reduce the likelihood of the holes being aligned.

2.7 Systems Thinking

Systems thinking seeks to understand how the different components of a system interact to promote the occurrence of events. When incidents happen, it is common for managers to focus on factors beyond their direct control or external to the system they manage, such as the poor WSH attitude of workers and supervisors, excessive market competition, and unsafe industry norms. Such a mindset is contrary to a systems thinking mindset. As discussed by Meadows and Wright (2008), many undesirable events like war, drug addiction, chronic disease, and environmental degradation, are unintended, but they continue to happen. Similarly, WSH incidents are unintended and they persist. These problems will only be eradicated when managers "stop casting blame, see the system as the source of its own problems, and find the courage and wisdom to restructure it" (Meadows and Wright, 2008, p. 4). Therefore, systems thinking is very much aligned with systemic incident causation models such as LCM and SCM.

However, systemic incident causation models such as LCM and SCM do not emphasise the importance of understanding the interactions between different parts of the organisation, such as management system, culture and leadership. These interactions generate complex system behaviours and unintended events. This is where systems thinking is useful for generating insights and suggestions for dealing with different situations.

It is more obvious that complexity can arise because of the unexpected interactions between physical components or production processes in high risk systems such as a nuclear plant or a pharmaceutical plant. This is what Perrow (2011) focuses on when he highlights the importance of understanding how *system accidents* can happen because of "unanticipated interaction of multiple failures" (p. 70). It is argued that WSH incidents are almost always complex; even the so-called straightforward incidents, e.g. fall from height and struck by falling object, have significant complexities as one delves into the causes of the incident.

A system, in a more general sense, is a set of elements that interact with each other for a specific purpose. In the case of WSH management, the WSH management system includes the documented policies, procedures and structures. The WSH management system also interacts with other management systems of the organisation, which have their own system elements. These management system elements interact with each other, and with the organisational culture (of which safety culture is a subset) and leadership to influence WSH-related behaviours, and hence the occurrence or non-occurrence of WSH incidents. The interactions occur over time and the cause and effect of the interactions and their subsequent influence on workers' behaviours are known as *dynamic complexity* (Senge, 2006), which are harder to comprehend, detect and mitigate. In contrast, *detail complexity* arises due to the large number of details and/or volume of information, which are comparatively more manageable using a *divide and conquer* approach. Systemic incident causation models like LCM and SCM are useful in identifying the underlying management, cultural and leadership components that contributed to the incident (i.e. its detail complexity), but systems thinking is more suitable for surfacing the dynamic complexity that gave rise to an incident.

2.8 Causal Loop Diagrams

The discipline of systems thinking has a range of tools for uncovering dynamic complexities of a system. A causal loop diagram (CLD) is one of such tools (Senge, 2006). Goh *et al.* (2010) introduced the use of CLDs in analysing incidents. In contrast to the linear nature of most incident causation models, a CLD can be used to describe the circular nature of cause and effect, and better explain the system's behaviour over time.

A CLD is made up of three basic processes: a reinforcing feedback loop, balancing feedback and delays. A reinforcing loop exists where a behaviour encourages similar behaviour in the future. It amplifies the behaviour across time and as the reinforcing loop continues, an accelerating growth or decline will occur. Figure 2.7 is an example of a reinforcing loop where the safe behaviour and its positive consequences, such as recognition by a supervisor and incentives from the company, create a virtuous cycle that encourages growth in the safe behaviour in the future, which then leads to further safe behaviour and positive consequences. A reinforcing loop can also be a vicious cycle where the unsafe behaviour (e.g. not doing safety checks) results in positive consequences like approval from peers and compliments from "unenlightened" clients who want to avoid

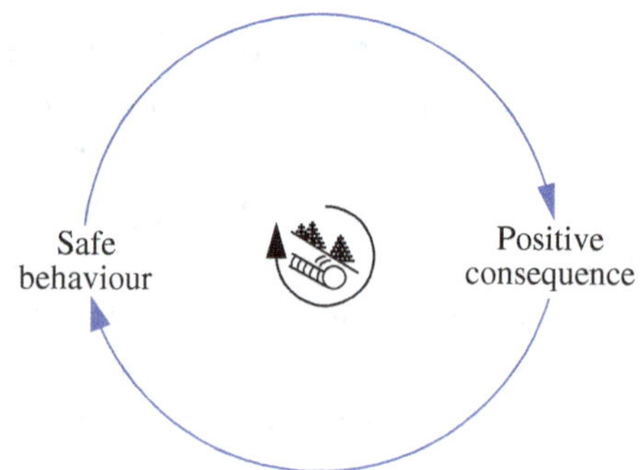

Figure 2.7 Example of a reinforcing loop

delays due to additional safety checks. These positive consequences will lead to more unsafe behaviour in the future.

Balancing feedback loops describe processes that aim to balance a behaviour or indicator at a target level. For example, as depicted in Figure 2.8, a company may be monitoring a safety performance indicator ("Actual level") and the company is motivated to increase its improvement effort whenever the target level is higher than the actual level. The larger the size of the gap, the greater the effort to improve the situation. When the size of the gap reduces, the pressure to improve the safety performance reduces. The way to sustain the improvement effort is to adjust the target level higher to maintain a healthy gap between target and actual levels. The target level need not be explicit and can be implicit and hidden in people's minds. It is also possible for both explicit and implicit target levels to exist, and people may unknowingly be focused on the implicit target despite the explicit target level being higher.

The subsequent effect of a cause may be delayed. With reference to Figure 2.9, when a safety behaviour leads to a positive consequence, it might take some time for workers to be convinced that the safe behaviour can indeed lead to positive

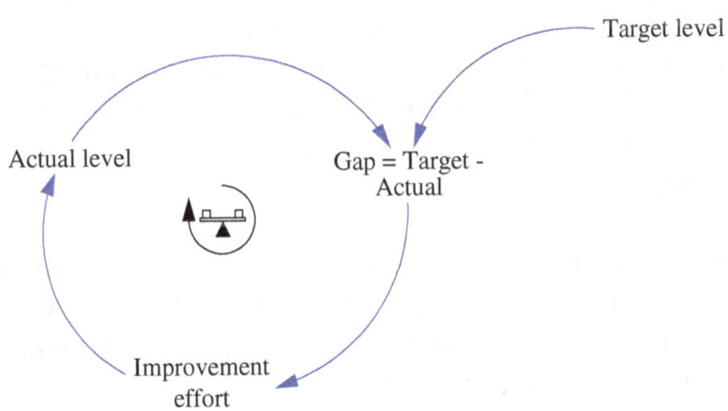

Figure 2.8   Example of a balancing loop

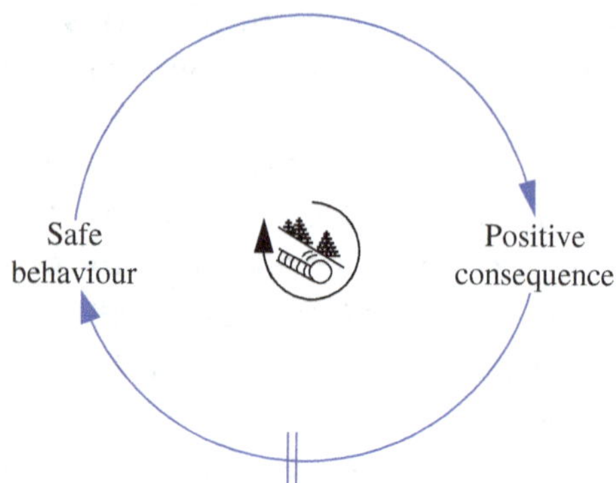

Figure 2.9   Example of a delay in a reinforcing loop

consequences. The delay is represented by two parallel lines on the arrow linking two variables. A lack of awareness of these delays can cause improvement efforts and safety programmes to be aborted prematurely.

Using the basic building blocks of reinforcing loops, balancing loops and delays, a series of systems archetypes can be captured to highlight common organisational problems and identify the possible leverage points that organisations can focus on to improve organisational performance. The systems archetypes are useful for understanding the problems inhibiting WSH management and provide suggestions on how to improve WSH performance. These archetypes are described in Senge (2006) and The Systems Thinker (2018). The following paragraph will describe the "shift the burden" archetype (see Figure 2.10) as an illustration of system archetypes. Other systems archetypes will be discussed in Chapter 3.

The archetype consists of two balancing loops, one addressing the symptoms, the other the underlying or fundamental causes of the problem. Symptomatic "solutions" can be very attractive to management, producing relatively quick positive results that focus on the symptoms and relieve the immediate pressure of the problem. The second balancing

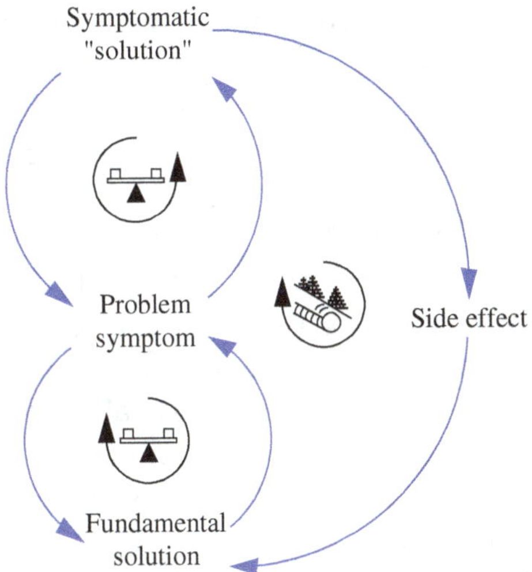

Figure 2.10   Shifting the burden archetype

process focuses on fundamental solutions to the problem that are more sustainable but that also have a delay between implementation and results. Failure to recognise this delay contributes to the tendency to focus on symptomatic solutions. When resources are focused on symptomatic solutions, the fundamental problem and the capacity of the organisation to resolve the problem worsens (side effect). Systems thinking concludes that while short-term efforts aimed at the symptoms can be useful, particularly to relieve the pressure as the delay between the fundamental solution and a manifestation of its benefits is experienced, but long-term solutions to a problem must focus on the fundamental loop.

Archetypes help managers understand the possible underlying factors sustaining the pattern of system behaviour, which leads to the occurrence of the incident. The archetypes also present leverage points and possible strategies that managers can use to improve their current situation. The archetypes and CLDs also allow managers to present their mental models to allow others to examine them and discuss solutions.

However, CLDs get too complicated very quickly. A well-known example is the CLD developed to describe the complexity of American military strategy in Afghanistan (Bumiller, 2010). General Stanley A. McChrystal, the leader of American and NATO forces in Afghanistan, remarked, "When we understand that slide (*with the CLD*), we'll have won the war". From a communication point of view, CLDs are only useful if the audience had been involved in the development of the CLDs. It is a communication tool for a team to discuss and present members' mental models of the situation they are evaluating. Furthermore, developing a useful CLD takes significant amount of skill and it is arguably more of an art than a science. System dynamics simulation is another example of a systems thinking tool that cannot be easily used in practice. The simulation model quantifies all variables and requires statistical and mathematical knowledge that escapes most managers.

As highlighted in Chapter 1, this book will introduce the application of the philosophy and core concepts of systems thinking in WSH management, and to avoid confusing the reader, more advanced tools such as CLDs and system dynamics simulation will not be discussed in detail.

2.9 Conclusions

This chapter presented several incident causation models, and it was established that systemic models are now widely adopted by organisations. Systemic models emphasise that organisational factors like the management system, leadership and safety culture are the fundamental causes of workplace incidents. However, they do not emphasise the interactions between system elements, which give rise to dynamic complexity which are more difficult for managers to identify and mitigate proactively. Thus, systems thinking was introduced as a useful management approach that can be used to improve WSH management systems. The following chapters will describe different WSH management topics and relevant concepts of systems thinking will be highlighted to help readers understand how systems thinking concepts can be applied in WSH management.

Review Questions

1. Describe the difference between an "event", an "incident" and an "accident".
2. Why is the Domino Theory no longer accepted in modern WSH management?
3. Give examples of events, patterns and underlying factors. Demonstrate how the underlying factors cause the patterns and then the events.
4. What is the key difference between a systemic model (e.g. Swiss Cheese Model) and non-systemic model (e.g. Energy Transfer Model)?
5. Explain the difference between immediate cause, basic cause and lack of control.
6. Explain the similarities and differences between the Swiss Chees Model and the Loss Causation Model.
7. Explain how detail complexity and dynamic complexity are different.
8. Explain how systems thinking can improve WSH management.
9. Search the internet for the different system archetypes. Explain how to select suitable archetypes for different system problems.

References

Bird, F. E., Germain, G. L., and Clark, M. D. (2003). *Practical loss control leadership*, Det Norske Veritas (U.S.A.), Inc., Duluth, Georgia.

Bumiller, E. (2010). *We Have Met the Enemy and He Is PowerPoint*. Retrieved from https://www.nytimes.com/2010/04/27/world/27powerpoint.html.

Christopher Tan (2017). "PIE uncompleted highway structure collapse: Cause likely to be human error, says veteran engineer." *Straits Times*, SPH, Singapore.

Chua, D. K. H., and Goh, Y. M. (2004). "Incident causation model for improving feedback of safety knowledge." *J. Constr. Eng. and Manage. — Am. Soc. of Civ. Eng.*, 130(4), 542–551.

Goh, Y. M., Brown, H., and Spickett, J. (2010). "Applying systems thinking concepts in the analysis of major incidents and safety culture." *Safety Science*, 48, 302–309.

Haddon Jr, W. (1973). "Energy damage and the ten countermeasure strategies." *Human Factors*, 15(4), 355–366.

Heinrich, H. W. (1936). *Industrial Accident Prevention*, McGraw Hill, New York.

Johnson, W. G. (1980). *MORT Safety Assurance System*, Marcel Dekker, New York.

Meadows, D. H., and Wright, D. (2008). *Thinking in systems: A primer.* White River Junction, Vt: Chelsea Green Pub.

Perrow, C. (2011). *Normal accidents: Living with high risk technologies-Updated edition.* Princeton university press.

Reason, J. (1997). *Managing the risks of organizational accidents,* Ashgate, Aldershot.

Sanders, M. S., and Shaw, B. (1988). *Research to determine the contribution of system factors in the occurrence of underground injury accidents,* Bureau of Mines, Pittsburgh.

Senge, P. (2006). *The fifth discipline — The art & practice of the learning organisation,* Random House Australia, New South Wales.

The Systems Thinker. (2018). Archetypes. Retrieved from https://thesystemsthinker.com/topics/archetypes/

Chapter 3

Incident Investigation

3.1 Introduction

This chapter will provide an overview of incident investigation. Incident investigation is an essential part of any workplace because it is important to determine the causes and underlying factors that led to the incident, so that similar incidents can be prevented from happening and the management system can become more effective in preventing incidents. The chapter will cover the purpose of incident investigation, an overview of a typical investigation process, types of evidence, incident analysis, the event causation technique (ECT) and how systems thinking concepts can be used to improve incident investigation.

3.2 Purpose of Investigation

According to ISO 45001:2018 clause 10.2 Incident, nonconformity and corrective action, organisations must "establish, implement and maintain a process(es), including reporting, investigating and taking action, to determine and manage incidents and nonconformities". In general, the aims of an investigation are to:

1. Determine root causes (similar to basic causes and latent conditions, as discussed in Chapter 2) and other factors that might be causing or contributing to the occurrence of incidents;
2. Determine if similar incidents have occurred or could potentially occur;

3. Review existing risk assessment (RA) and determine and implement any action needed, including corrective actions (actions to eliminate the causes of a detected nonconformity (i.e. non-fulfilment of a requirement) or other undesirable situation, and to prevent recurrence), in accordance to the hierarchy of controls and the management of change;

4. Recommend changes to the workplace safety and health (WSH) management system, if necessary; and

5. Communicate the investigation results to workers and other interested parties.

ISO 45001:2018 emphasises the importance of workers' participation and involvement of other relevant interested parties (e.g. contractors and suppliers) during an investigation. One of the key reasons is that the corrective actions developed will have to be implemented by the workers or the relevant interested parties, and it is critical for them to be involved in the development of corrective actions arising from the incident investigation. This will better ensure the feasibility and effectiveness of the corrective actions.

During an investigation, it is common for the investigation team to identify opportunities for improvement that may not be directly related to the incident. These opportunities for improvement could be possible causes of the incident which were proven to be unrelated to the incident. For example, when a team of investigators were investigating an incident involving a worker who tripped over a wire in a kitchen, they might realise that although the old tiles used in the kitchen did not contribute to the fall, non-slip tiles would help to reduce the risk of slips in the kitchen in the future.

According to ISO 45001:2018, a continual improvement activity is a "recurring activity to enhance performance". The activity can be based on the Plan–Do–Check–Act (PDCA) cycle and are typically not directly arising from the incident investigation per se.

It should be noted that an evaluation of opportunities for improvement and continual improvement activities can occur at any time and need not be triggered by an incident. However, a thorough investigation process would typically be able to identify multiple opportunities for improvement. See Case Study 1 for examples of corrective action, opportunities for improvement, and actions for continual improvement.

Case Study 1 — Forklift Accident

A worker was struck by a forklift at a factory leading to permanent injuries to the legs of the injured worker. The investigation found that the forklift driver was driving the forklift forward while carrying a load, which blocked the driver's view. The safe method was to reverse the forklift to ensure that there was a clear view of the path and any pedestrians in the vicinity. During the investigation it was discovered that the worker who drove the forklift did not attend suitable forklift training and had driven the forklift without permission. However, the forklift key was in the forklift and was not placed in the key press as per procedure. A key press is a small cupboard meant to control access to keys through a log book system. There was no clear requirement on how frequently the supervisor, manager and workplace safety and health officer (WSHO) overseeing the key press should check on the keys in the key press.

Causes: Blocked vision when driving forklift ← (was caused by) unsafe driving ← untrained forklift driver ← uncontrolled access to forklift keys ← key not placed in key press ← inadequate monitoring and control of key press

Corrective actions for "inadequate monitoring and control of key press"

1. Develop a procedure for control of the key press, where all drivers must return the keys to the key press whenever the forklifts are not in use. Supervisors, WSHOs and managers have to conduct different levels of checks as stipulated in the new procedure.

(*Continued*)

Case Study 1:   (*Continued*)

Corrective actions for "key not placed in key press"

1. Make missing keys obvious with good visual markers in the key press.
2. Drivers will be required to place an ID at the key press in exchange for the key.
3. All forklift drivers are briefed on the importance of removing the keys from the forklift and returning the keys to the key press. Disciplinary action will be taken if drivers were found to flout the safety rule. Incentives will be provided for drivers that comply with the procedure consistently.

Corrective actions for "untrained forklift driver"

1. All workers are reminded that unauthorised use of the forklift is dangerous and will be punished.
2. Authorised forklift drivers will be identified with a brown helmet with a name on it. Pictures of authorised forklift drivers are displayed on the notice board.

Opportunities for improvement arising from this accident

1. Preventive maintenance of the forklift will be implemented.
2. Additional pedestrian pathways will be created to reduce the likelihood of collision.

Actions for continual improvement arising from this accident

1. The effectiveness of the measures will be reviewed by the safety committee 1, 3 and 6 months after implementation. Subsequently, a review will be conducted every 6 months as part of the biannual audit.

The key purpose of an investigation is to prevent a recurrence, and improve the management system and culture by eliminating or mitigating the systematic patterns of behaviour and management that produced the incident. As discussed in Chapter 2, incidents can be very costly. Thus, it is important that investigations go deep enough to identify patterns and underlying factors and not stay on the event level. This means that instead of simply blaming or even firing (a reactive event level response) the workers involved in an incident for violating the safety rules, the investigation team should consider how the unsafe behaviour was facilitated by the situation that the workers were in (a proactive system level response). That said, it must be noted that disciplinary actions are still necessary to deter unsafe behaviours, but investigations should not focus on unsafe acts and punishments only.

Figure 3.1 demonstrates how incident investigation fits into the overall picture of WSH management. Incident investigation is a reactive component of a management system. It is reactive in the sense that an investigation is triggered only when there is an incident, which is undesirable due to its consequences or potential consequences. Nevertheless, incident investigations serve an important role in improving the way operations are being managed and controlled. On the

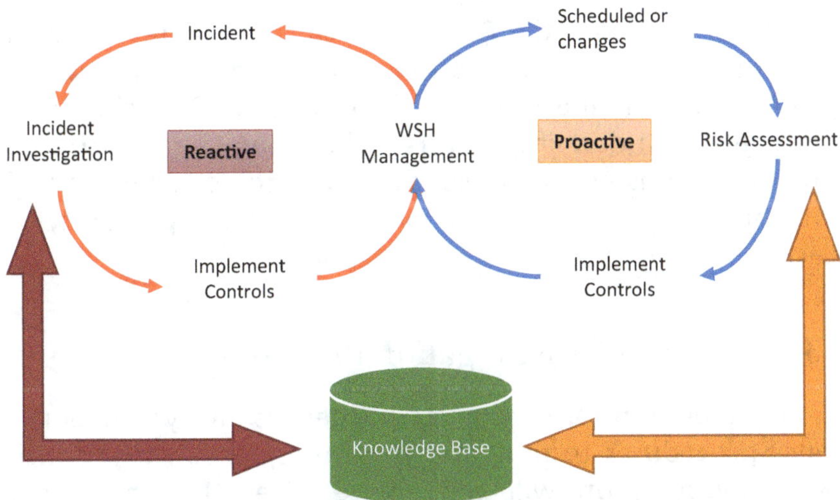

Figure 3.1   Dependence between reactive and proactive control loops; adapted from Chua and Goh (2004)

other hand, the ideal situation would be for hazards and risk controls to be identified proactively so that operations are safe and incident-free. The key underlying process in WSH management that promotes proactive management is RA. RAs are conducted prior to operations, at scheduled periods, after an incident, when there are changes in the operations (e.g. method, equipment and material), and when there are new WSH information that justifies a review. As depicted in Figure 3.1, both proactive and reactive loops aim to improve the WSH performance of the operation.

To ensure that the organisation learns from its past experiences, it is important to capture the information generated during incident investigation and RA in a knowledge base, so that investigators and RA (or risk management (RM)) teams can access the knowledge base when they are conducting investigations and RAs respectively. For example, during an investigation, it is important to consider the risk controls that were identified in the RA. The investigators should assess if the risk controls were comprehensive, suitable, and implemented. In this way, the RA process can also be reviewed as part of the incident investigation process. An incident investigation should make full use of the undesired incident to improve the underlying factors influencing the patterns of how operations are being conducted in the organisation. If not, the organisation would have missed an opportunity to learn.

3.3 Overview of Investigation Process

The immediate priority after an incident is always the emergency procedures to mitigate the consequences of the incident. Investigation will take place after the emergency response phase is over. Nevertheless, whenever possible, efforts should be taken to preserve the incident scene (e.g. restricting access to the incident scene and ensuring that objects related to incident are left untouched) so as to facilitate the subsequent investigation process. It is important to conduct a simple RA prior to the emergency response to identify possible hazards to rescuers and provide necessary controls to assure their safety and health during the rescue. For example, prior to the rescue it is important to ensure that all machinery have been shut down, electrical power isolated, the safety data sheet of the chemicals reviewed, and suitable personal protective equipment provided.

An investigation can be split into the four inter-related phases: (1) notify, (2) investigate, (3) analyse and (4) report. Organisations need to be notified of incidents before an investigation can be conducted. During notification, the organisation needs to prioritise and allocate resources to the investigation. Some considerations include the severity of the incident, the likelihood of a recurrence, the worst credible consequence of the incident, and the resources available. Many organisations launch investigations only when the incident severity is high, e.g. when a worker is hospitalised or severely injured. However, the actual severity of an incident is frequently a matter of luck. A worker struck by a falling wrench could have easily survived if the wrench fell one second later, and that will reduce the incident consequence to zero. Thus, organisations should assess the worst credible consequence of an incident, and conduct an investigation if the severity of the worst credible consequence is high. It is also important for the organisation to communicate with key stakeholders, e.g. government agencies, clients, workers and unions, on the investigation's purpose and process so as to allay concerns and garner support for the investigation. A typical incident notification form is shown in Figure 3.2.

The investigation phase is a fact-gathering process which focuses on obtaining the facts and information related to the incident. The fact-gathering process is intertwined with the

Incident Notification Form

Date of Notification:	Time of Notification:
Date of Incident:	Time of Incident:
Location of Incident:	Type of Incident (injury/property loss/ill health/near hit/others):
Severity of Injury/Ill health:	
Potential Severity of Injury/Ill health (worst credible):	Activity Prior to Incident:
Likelihood of Recurrence of Incident:	High-Potential Incident? (Yes/No)
Type of Personnel Injured/Diseased:	Name and ID of Personnel Injured/Diseased:

Incident Description (attach photos and sketches):

Actions Taken:

Subsequent Actions:

Submitted by:

Figure 3.2.   Example of an incident notification form

"analyse" phase, which includes hypothesising about what happened, causes, and underlying factors, and evaluating the hypotheses against the facts gathered. The analysis phase aims to identify the deep-rooted causes and underlying factors of the incident so as to prevent the recurrence of incidents. The analysis process can involve the use of different methods and tools to guide investigators to probe into possible causes. Missing facts and evidence will be identified during the analysis and hence guide the investigation. As more facts and information are gathered, the hypotheses about the incident and its causes and underlying factors will be confirmed, amended, or eliminated. New hypotheses can also be developed during the analysis. Thus, investigation and analysis activities are inter-related.

Based on the author's experience, the competency and mindset of the investigators, and the management's commitment towards uncovering underlying factors, are more critical than the incident analysis methods. A sophisticated tool or method, e.g. AcciMap and Functional Resonance Accident Model (FRAM), is useless in the hands of people who are not competent or not committed. Asking a series of "whys" is perhaps the most basic form of analysis and it can be extremely effective if the investigators know how to frame the questions, and the management is determined to uncover the systemic issues influencing the incident. However, if the investigators are competent, an incident analysis method can help to make the investigation more structured and comprehensive. This chapter will provide a brief review of some of the established incident analysis techniques.

Finally, the reporting phase is about presenting the findings and recommendations to management and other stakeholders. A report is usually written and submitted to the management for discussion and approval. Many organisations use standard templates for investigation reports. An example is shown in Figure 3.3, which comes with a set of basic incident investigation

Incident investigation form

Incident details			
Name of person involved in the incident:		Date of incident:	
Location of incident:			

Incident investigation team:

What task was being performed at the time of the incident?

What happened? (e.g. 'employee tripped over box' or 'forklift hit wall')

What factors contributed to the incident?

Environment:		Equipment/materials:	
☐ Noise	☐ Layout / design	☐ Wrong equipment for the job	☐ Equipment failure
☐ Lighting	☐ Dust / fume	☐ Inadequate maintenance	☐ Material / equipment too heavy / awkward
☐ Vibration	☐ Slip / trip hazard	☐ Inadequate guarding	☐ Inadequate training provided
☐ Damaged / unstable floor	☐ Other	☐ Other	
Work systems:		**People:**	
☐ Hazard not identified	☐ No / inadequate risk assessment conducted	☐ Procedure not followed / no procedure exists	☐ Drugs / alcohol
☐ No / inadequate safe work procedure	☐ No / inadequate controls implemented	☐ Fatigue	☐ Time / production pressures
☐ Hazard not reported	☐ Inadequate training / supervision	☐ Change of routine	☐ Distraction / personal issues / stress
☐ Other		☐ Lack of communication	☐ Other

Corrective actions:

Contributing factor (from above list)	What are we going to do to fix the problem?	Who	When	Completion date

Issue fixed?			
Name		**Signature**	**Date**
Person involved in incident:			
Manager:			

Figure 3.3.   (*Continued*)

Incident investigation process guide

1. Establish the facts of the incident, including:
    - What happened?
    - When and where did it happen?
    - What task was being done?
    - Who was involved?
    - Were there any witnesses?

2. Gather all necessary background information, for example:
    - maintenance records
    - safe work procedures
    - instructions manuals
    - training records.

3. Consider all the potential contributing factors:
    - Environment: *Did environmental conditions (e.g. light, noise, floor surfaces) contribute to the incident?*
    - Equipment /materials: *Did anything about the equipment, materials, tools etc (e.g. equipment failures, missing guards) contribute to the incident?*
    - Work systems: *Was there something about the system that contributed (e.g. hazard not identified, known hazard not addressed)?*
    - People: *Was there something the workers, supervisors or contractors did that contributed to the incident (e.g. poor communication, being tired or rushing to finish on time)?*

4. Determine the primary cause/s of the incident, that is, those which if they hadn't occurred then the incident wouldn't have occurred. Ask yourself *"Would the incident have happened if....?"*

5. Identify the root cause / system failures that underlie the primary cause/s and contributing factors.

    One simple technique for identifying the root cause is the 'Five Whys'. This technique involves asking yourself 'Why did this happen?' and continuing to ask 'Why' for each response until you reach a conclusion that does not generate another 'why' and the underlying cause becomes apparent.

6. The final and most import step in any investigation is to take action to fix all the factors that contributed to the incident, starting with the primary cause/s and working through each of the contributing and underlying causes.

Figure 3.3.   Incident investigation form by WorkCover Queensland (n.d.)

guidelines (Incident Investigation Form (WorkCover Queensland, n.d.)). As shown in the form in Figure 3.3, the report provides the basic information about the incident, the injured and the task being performed at the time of the incident. More importantly, the report must contain the causes and underlying or contributory factors and the corrective actions. The corrective actions should be tracked to ensure that they are implemented. A more detailed report will be needed if the incident is more complex, e.g. involves multiple parties, multiple machinery, and/or a complicated series of events.

It is important to develop and maintain a set of systematic and comprehensive investigation policies, procedures and templates. This includes a policy statement demonstrating clear management emphasis on the need to prevent the recurrence of incidents through thorough incident investigations. The policy statement should also be aligned with a systemic incident causation model by focusing incident investigations on fundamental management issues related to the organisation's management system, culture and leadership. Investigation resources, procedures and supporting forms and templates such as notification forms, investigation guidelines, a list of investigators for different types of accidents, pre-identified competencies required for different investigations, and pre-allocated resources for investigators (e.g. an investigation kit, financial resources, and time). It is useful to involve operations staff in investigations as they are the most familiar with the operations. Workplace safety and health professionals who are more familiar with the WSH legislations and investigation processes can assist the operational staff and advise them on the investigation's purposes and processes. These roles and responsibilities must be pre-determined as part of the investigation procedure. The investigation guidelines provided by WSH regulatory bodies (e.g. Workplace Safety and Health Council (2013) and Health and Safety Executive (2004)) are useful guidance for investigation procedures.

It should be noted that the investigation process is not an exact science. Many-a-times, the investigation team will need to use their judgement and opinions to assess the facts and evidence gathered. However, it must be clearly stated when something is an opinion or a judgement, and the team should not mislead readers of the report into thinking that these opinions or judgements are factual. In addition, it is important for investigators to realise that they are an observer of an event that happened in the past. This retrospective view allows the investigator to see the hazards and the consequences of decisions more easily. However, the "actors" or people involved in the incident were looking forward as the events unfolded. Furthermore, the actors are busy people who are always juggling safety, productivity, quality and other issues including personal problems in life. The actors do not have full information about their work environment, and are also uncertain about the consequences of their actions and other people's actions. According to Dekker (2017), this is similar to walking in a tunnel, which limits the view of each actor. The individual characteristics of these actors, like mindfulness, diligence and capability, are important factors, but investigators need to look beyond the individuals to consider the reasons for the mistakes and errors that the individuals made. Knowing the reasons for human errors is critical in preventing a recurrence because a different person placed in the same situation can easily make the same error.

For example, in Case Study 1, the unauthorised forklift operator who took the key without authorisation was probably balancing different work pressures while trying to make the best use of the time and resources that he was given. Even though it is still necessary to provide deterrence (i.e. punishment) for workers who commit such violations, it is more effective to prevent unauthorised access to the keys and reduce the likelihood for such violations. In contrast, managers can blame the operator for violating the safety rules, punish him and assume that other workers will not violate said safety rules again after witnessing a worker being punished. Most of the time, such punishment will result in fear in the organisation and can cause workers to stop providing WSH-related information to management. However, this does not mean that a "no-blame" policy will work. There is a need for fair punishment, which takes into account both the actual challenges that individuals face, and focuses on developing systems that prevent mistakes and violations. The need for a fair or "just culture" will be discussed in Chapter 7.

3.4 Types of Evidence

There are four main types of evidence: part, position, people and paper. "Part" refers to evidence in the form of material, equipment and parts of the environment such as the worn-out tyres of a forklift, the ignition key of a motor vehicle, the dust level or heaps of dust, the noise level, the illumination level, the braces of scaffolds, and lanyards. Part evidence is usually collected from the incident scene. One of the first actions of an investigation is to cordon off the incident scene to prevent part evidence from being removed or altered. "Position" refers to the physical relationship between the part(s) and people involved, and the placement of the evidence at different points of time. Some examples include the position of the injured before, during and after the incident, the location of the safety data sheet (for chemicals or hazardous substances) before and during the accident, the position of the wheel of the truck after the explosion, and the position of the crane after the collapse.

"People" evidence refers to the statements provided by people who have information related to the incident. Some examples include interviews with the injured, witnesses (people that saw the incident), people associated with the activity and people with information relevant to the investigation. Besides direct eye-witnesses and people directly involved in the activity, investigators frequently need to interview suppliers of machinery, experts or experienced workers who have a good understanding of how work is supposed to be conducted, and managers who are accountable for the work area in general. Interviews with witnesses and people with direct information about the incident should be conducted as early as possible because people's memories tend to distort with time. In addition, witnesses might become affected by other people's intentional or unintentional comments. Investigators must also obtain "papers" and documents (including electronic and video footage) related to the incident. These documents include, for example, standard operating procedures or method statements, job orders, emails, close circuit TV footage, safety data sheets (SDS), safety committee meeting minutes, shift change log books, permits-to-work, data logger records, and RAs.

The investigation team will have to decide what evidence needs to be collected. This is usually based on investigation procedure, investigators' experience, the hypotheses of the incident, and the analyses that the investigators conduct iteratively. Each cause should ideally be based on two or more pieces of evidence.

3.5 Overview of Incident Analysis Methods and Techniques

There are many incident analysis methods and techniques available in the literature. The Energy Institute (2008) captured 28 incident analysis methods that are suitable for analysing underlying (human and organisational) factors, but many methods were excluded from the document (e.g. AcciMap, Systems-Theoretic Accident Model and Processes (STAMP) and Man–Technology–Organization (MTO) analysis (Lappalainen and Perttula, 2017)). Most established incident analysis methods are useful in making incident investigation more thorough and structured, but as indicated earlier, their degree of usefulness really depends on how the investigator makes use of the method and whether the management is committed to uncovering systemic issues.

Table A2 in the guidance provided by The Energy Institute (2008) differentiated 28 incident investigation methods based on the following features:

1. Whether training is required
2. Whether a software is required
3. Whether the method is designed to retrospectively evaluate existing investigation reports
4. Whether the method is used in the petroleum industry (*or the relevant industry*)
5. Whether the method requires the development of a pictorial representation of the incident
6. Whether the method is dependent on other tools to complete the analysis
7. Whether the method provides standard corrective actions and solutions
8. Whether the method provides checklists or flow charts to guide the analysis

These are important consideration when an organisation is selecting an incident investigation method. It should be noted that standard correction actions and solutions are usually not very useful because they are usually too general or basic.

Even though there are differences between the incident analysis methods, they also have significant similarities. For example, most of the techniques are guided by systemic causation models such as the Loss Causation Model (LCM) and the Swiss Cheese Model (SCM), which emphasise the importance of uncovering underlying factors, systemic issues or root causes. SCM is especially influential and several methods such as bow-tie, the Human Factors Analysis and Classification System (HFACS), and Incident Cause Analysis Method (ICAM) are based on the model. Another similarity across the techniques is the concept of multi-causation, where each event or cause arises due to a combination of multiple events and causes. Several methods emphasise the evaluation of "barrier analysis" (see discussion on SCM in Chapter 2), where the target (people or property that can be harmed), barrier (control measures that prevent the target from being harmed) and energy (harmful energy or substance that can harm the target) have to be identified. Furthermore, a common feature of the investigation methods is the use of a timeline or chronological sequence of events. Last, each method typically has a set of prompts that guide the investigator in identifying possible events, causes and underlying factors.

The author recommends the use of incident analysis methods that give experienced investigators flexibility in conducting their investigation and provide less experienced investigators enough guidance to conduct their initial cases. Software-based methods can incur additional costs and if the investigators are not used to the software, it may impede the brainstorming of possible hypotheses about the incident. Whiteboards, post-its and sketches are usually more useful during brainstorming and discussions, after which, information can be summarised in typical presentation, spreadsheet or word-processing software. The key features of a useful incident analysis method are summarised below:

1. Flexible and adaptable to the preferences and needs of the investigators
2. Requires investigators to develop a chronological sequence of events
3. Emphasises the concept of multi-causation, where investigators are encouraged to identify a comprehensive set of necessary and sufficient conditions for the incident to occur
4. Prompts investigators to evaluate control measures that were or should have been identified prior to the incident
5. Reminds investigators to uncover systemic issues (underlying factors) which contributed to the incident

This book introduces the Event Causation Technique (ECT), which was created based on the features described above.

3.6 Event Causation Technique

The ECT is depicted in Figure 3.4 and Table 3.1 shows the corresponding taxonomy, which is not exhaustive. The ECT is based on the PhD dissertation of the author, and the technique was refined across the years after applications in actual accident investigations and analyses. The technique is an extension of incident causation models like the Energy Transfer Model (ETM), the LCM and the SCM, but unlike these models, the ECT is meant to be a structured, simple and flexible step-by-step incident analysis technique. Investigators and managers using the ECT should be able to determine a comprehensive set of causes and factors so as to recommend actions to prevent the recurrence of an incident. It is also important to note that ECT is meant to provide feedback to the proactive RA process (see Figure 3.1), where the investigators are expected to review the control measures identified in the RAs relevant to the incident.

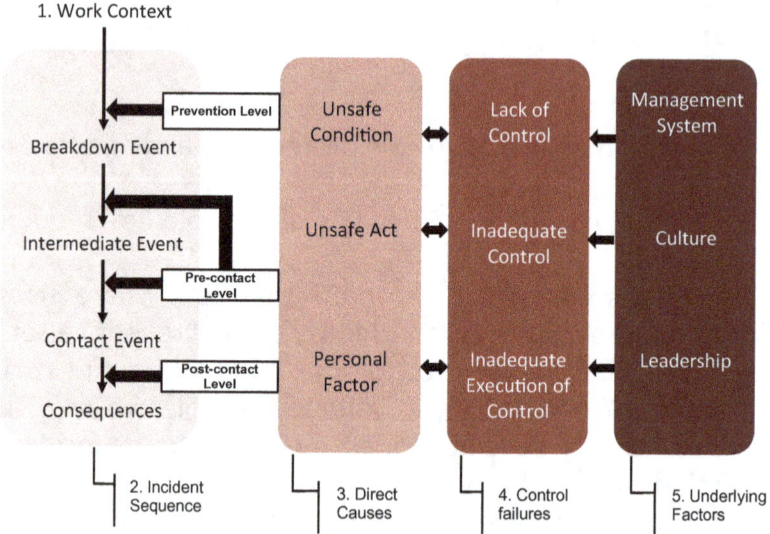

Figure 3.4 Event causation technique version 2

Table 3.1   Event causation technique taxonomy

Consequences		
1. Injured worker time	2. Co-worker time	3. Leader time
4. General losses (lost production time, decreased effectiveness of employees and loss of business goodwill, etc.)	5. Property losses	6. Other consequences
Incident Events — Breakdown, intermediate and contact events		
1. Collapse of structure	2. Loss of balance	3. Release of (any harmful energy or substance)
4. Loss control of plant/equipment	5. Overstress/ overexertion/ overload	6. Fall to lower level (either the body falls, or the object falls and hits the body)
7. Fall on same level (slip and fall, trip over)	8. Struck by (hit by moving object)	9. Struck against (running or bumping into)
10. Contact with (any harmful energy or substance)	11. Caught in (pinch and nip points)	12. Caught on (snagged, hung)
13. Caught between (crushed or amputated)	14. Other events	
Direct Causes		
Unsafe Acts and Practices		
1. Operating equipment without authority	2. Failure to warn	3. Failure to secure

(Continued)

Table 3.1   (*Continued*)

Direct Causes		
Unsafe Acts and Practices		
4. Operating at improper speed	5. Making safety devices inoperable	6. Using defective equipment
7. Using equipment improperly	8. Failing to use personal protective equipment properly	9. Improper loading
10. Improper placement	11. Improper lifting	12. Improper position for task
13. Servicing equipment in operation	14. Horseplay	15. Under influence of alcohol and/ or other drugs
16. Failure to follow procedure/ policy/ practice	17. Failure to identify hazard/risk	18. Failure to check/monitor
19. Failure to react/ correct	20. Failure to communicate/ coordinate	
Unsafe Conditions		
1. Inadequate guards or barriers	2. Inadequate or improper protective equipment	3. Defective tools, equipment or materials
4. Congestion or restricted action	5. Inadequate warning systems	6. Fire or explosion hazards
7. Poor housekeeping; disorder	8. Hazardous environmental conditions: gases, dusts, smokes, fumes, vapours	9. Noise exposure
10. Radiation exposure	11. Temperature extremes	12. Inadequate or excess illumination

(*Continued*)

Table 3.1   (*Continued*)

Unsafe Conditions		
13. Inadequate ventilation	14. Inadequate instructions/ procedures	15. Other unsafe condition(s)
Personal Factors		
1. Inadequate physical/ physiological capability	2. Inadequate mental/ psychological capability	3. Physical or physiological stress
4. Mental or psychological stress	5. Lack of knowledge	6. Lack of skill
7. Improper motivation		
Control Failures		
1. Failure to prevent the initial build-up of the energy or substance	2. Failure to reduce the amount of energy or substance build-up or stored	3. Failure to prevent the release of the energy or substance
4. Failure to reduce the rate of spatial distribution of release of energy or substance from its source	5. Failure to separate in space or time the energy or substance being released from the susceptible structure or person	6. Failure to separate the energy being released from the susceptible structure or person by interposition of an effective barrier
7. Failure to decrease the hazardous nature of contact surface, subsurface, or	8. Failure to strengthen the structure or person which might be	9. Failure to move rapidly in detection and evaluation of damage and to

(Continued)

Table 3.1 (*Continued*)

Control Failures		
basic structure which can come into contact with susceptible structure or person	damaged by the energy transfer or contact with substance	counter its continuation and extension
10. Failure of all measures which fall between the emergency period following the damaging energy exchange and the final stabilisation of the process	11. Failure to ensure that relevant employees have sufficient awareness, knowledge, and/ or skills	12. Failure to provide adequate supervision
13. Other control failures		
Underlying Factors		
Leadership (the buck stops here)		
1. Failure to take accountability	2. Failure to maintain oversight of WSH management system	3. Failure to maintain oversight of WSH culture
4. Failure to ensure adequate resources for WSH	5. Failure to communicate importance of WSH	6. Failure to protect workers who report incidents, hazards, risks, etc.
7. Lack of WSH knowledge	8. Lack of WSH leadership skills	9. Lack of commitment or inappropriate attitude or motivation towards WSH

(Continued)

Table 3.1   (*Continued*)

Underlying Factors		
Leadership (the buck stops here)		
10. Mental or psychological stress	11. Failure to demonstrate commitment to WSH	12. Other leadership factors
Management System Inadequacies (based on ISO 14001:2015)		
1. Policy	2. Organisational roles, responsibilities and authorities	3. Consultation and participation of workers
4. Risk assessment	5. Determination of legal and other requirements	6. Planning processes
7. Resource allocation	8. Competence	9. Awareness and communication
10. Documentation	11. Risk controls	12. Management of change
13. Procurement	14. Emergency preparedness and response	15. Monitoring, measurement, analysis and performance evaluation
16. Internal audit	17. Management review	18. Incident, nonconformity and corrective action
19. Continual improvement	20. Other management system elements	

(*Continued*)

Table 3.1   (*Continued*)

WSH Culture (Shared WSH values and beliefs in the organisation or at different organisational levels)		
1. WSH is given lower priority as compared to other organisational goals, e.g. productivity, deadlines and profit	2. Believes that incidents will not happen or insensitive to WSH hazards	3. Unsafe behavioural norms or practices
4. Other WSH culture issues		

The ECT has five main components: (1) work context, (2) incident sequence, (3) direct causes, (4) control failures and (5) underlying factors. The following sub-sections will explain each of these components in more detail.

3.6.1 Work context

Work context describes the workplace scenario where the incident took place. Descriptors of work context can include the type of work being executed, the types of trades and workers involved, the equipment and material used in the work, the work environment where the work took place and other interacting or nearby work. The descriptors can be used to categorise incident cases during incident data analysis. The work context, or parts of the work context, together with the direct causes are the prerequisites of an incident.

3.6.2 Incident sequence

Incident sequence is a concept adapted from the ETM described in the earlier chapter. It splits the incident into five types of key events: preceding event, breakdown event, intermediate event, contact event and consequence. The types of losses and events in the LCM are useful references for the incident sequence. The incident sequence describes what happened just before the incident till the occurrence of the consequences, so that opportunities for preventing future occurrence of the incident can be identified.

A breakdown event is defined as an initiating point of loss of control of a source of energy or substance that, without an intervening event (e.g. presence of a control), will lead to the occurrence of a contact event. In contrast, a contact event is an event where the victim comes into contact with the source of hazardous energy or substance. An intermediate event refers to any significant event between the breakdown event and contact event that, if adequately controlled, will prevent or mitigate the consequences of the incident. It is not necessary to include an intermediate event (i.e. it is optional), or there could be more than one intermediate event if the incident is complex. The key criteria for deciding if intermediate events should be captured in the incident sequence is if there are opportunities for controlling the intermediate event. If the intermediate event represents an opportunity to put in a control to prevent future occurrence of the incident, then it must be described in the incident sequence. On the other hand, an intermediate event may be included to ensure that the incident sequence is complete and logical to the reader, but it may not provide an opportunity for control. Consequences refer to the undesirable effects of the incident, e.g., property loss, environmental pollution, number of man-days lost and type of injury.

It is beneficial to define incidents based on the incident sequence structure, so that causal factors and controls can be classified systematically into three levels of intervention and causation, namely, "post-contact", "pre-contact" and "prevention" levels. By focusing on the three levels of intervention and causation, investigators will be encouraged to broaden their scope of investigation to identify more opportunities for improvement. For example, when a worker loses his or her balance and falls off the edge of a building, an investigator could easily state that the 'main cause' of the accident is that the worker was not using the fall arrest system provided. Even if the underlying factors and the controls that had contributed to the contact event (striking the ground) were identified, the recommendations based on the investigation would probably only prevent the recurrence of the contact event, but not the breakdown event (i.e. the worker's loss of balance, in this instance). To effectively lower the risk of the activity, the factors that contributed to the occurrence of the breakdown event, the intermediate event, the contact event and the consequences of the incident should all be identified.

The event just prior to the breakdown event, i.e. the preceding event, can be included to better describe the incident sequence. The preceding event can help to clarify the work context. It typically describes the actions being done by the workers involved in the incident prior to the breakdown event. The preceding event is not necessary if the work context provides sufficient information about what happened prior to the breakdown event. It should also be noted that the preceding event should not overlap with the information captured under direct causes, e.g. an unsafe act by a worker.

3.6.3 Direct causes

As in the case of "immediate causes" in the LCM, direct causes refer to causes that are directly linked to an event in the incident sequence. A direct cause is a necessary condition for an event to occur. When the direct cause is removed, the event will not occur. Direct causes are classified into unsafe conditions, unsafe acts and personal factors. Unsafe conditions refer to any direct cause related to non-human aspects, e.g., faulty equipment, unsuitable material and unsafe work environment. Unsafe acts refer to observable actions or lack of action of front-line personnel executing the work. In most instances, the definitions of safe and unsafe acts are dependent on RAs and relevant standards and legislations. In broad terms, a condition or action that leads to an unacceptable risk level is unsafe. When standards and legislations are applicable, non-compliance is usually assumed to define what is unsafe.

Personal factors refer to factors such as the physiological, mental and psychological factors of a person (Bird *et al.*, 2003). Some examples include inadequate strength, insufficient physical endurance, lack of knowledge and improper motivation. Within the context of direct causes, it is the personal factors of front-line workers or people directly involved in the incident that are considered. For example, consider a worker who performed an unsafe act by climbing up the bracings of an access scaffold. A common personal factor that led to this unsafe act is improper motivation to save time and effort.

It should be noted that direct causes may not be present if the cause of the incident event is a control failure. This means that the incident event directly links to the control failure.

3.6.4 Control failures

Control failures refer to failures of specific controls that an organisation has or should have. These controls should have been highlighted in an RA that the organisation conducted prior to the incident. Failure occurs when:

- There is a lack of control, i.e. control was not identified in the relevant RA,
- There is inadequate control, i.e. control was identified but it was not adequate in preventing the direct cause and/or incident event, or
- There is an inadequate execution of control, i.e. the control could have prevented the direct cause and/or incident event, but it was not implemented.

The possible control failures in Table 3.2, which were extracted from Table 3.1, are based on the ten counter-measures of the ETM (Haddon Jr, 1973). The relevant incident events, i.e. breakdown event, intermediate event, contact event, and consequence, and corresponding level of intervention and causation were also identified in the table. Relevant ECT events refer to the incident event that can occur if the control failure occurs. However, with a wide variety of possible incident sequence and control failures, Table 3.2 is only a guide and exceptions are expected in some cases. Some examples of controls are edge barricades, warning signs, training, personal protective equipment and emergency evacuation paths.

Controls are usually identified during an RA or RM process, which will be covered in Chapter 4. Controls need to be specific and must have direct relation to the direct cause and/or incident event. A lack of control refers to the absence of a control, for example, the absence of necessary barricades. Inadequate control refers to the presence of a control, wherein the design or planning of the control was not adequate. For example, a barricade designed with insufficient height, where the open edge becomes inadequately protected, hence resulting in an unsafe condition. Inadequate execution of control refers to non-compliance with the initial plan or design of a control. The unsafe act of using a non-explosion-proof torchlight in an area designated as a high fire or explosion risk zone can be due to an inadequately executed briefing (the control). In this example, the briefing could have failed to cover some of the required content, including the locations of the high fire-risk zones.

With reference to Figure 3.1, it is important for investigators to evaluate the RA and the relevant controls. This will help to improve the effectiveness of the RA process, which is the cornerstone of any WSH management system. The information about control failures uncovered during investigation will be

Table 3.2   Types of control failures

Possible Control Failures	Relevant ECT Event (Typical)	ECT Level of Intervention & Causation	Relevant Hierarchy of Control
1. Failure to prevent the initial build-up of the energy or substance	BE	Prevention	Elimination
2. Failure to reduce the amount of energy or substance build-up or stored	BE, CSQ	Prevention, post-contact	Substitution
3. Failure to prevent the release of the energy or substance	BE, IE	Prevention	Engineering, administrative
4. Failure to reduce the rate of spatial distribution of release of energy or substance from its source	IE, CE	Pre-contact	Substitution, engineering
5. Failure to separate in space or time the energy or substance being released from the susceptible structure or person	IE, CE	Pre-contact	Engineering, administrative
6. Failure to separate the energy being released from the susceptible structure or person by interposition of an effective barrier	IE, CE	Pre-contact	Engineering

(Continued)

Table 3.2 (*Continued*)

Possible Control Failures	Relevant ECT Event (Typical)	ECT Level of Intervention & Causation	Relevant Hierarchy of Control
7. Failure to decrease the hazardous nature of the contact surface, subsurface, or basic structure which can come into contact with susceptible structure or person	IE, CE	Pre-contact	Substitution
8. Failure to strengthen the structure or person which might be damaged by the energy transfer or contact with substance	IE, CE, CSQ	Pre-contact, post-contact	Engineering, personal protective equipment
9. Failure to move rapidly in detection and evaluation of damage and to counter its continuation and extension	CE, CSQ	Post-contact	Administrative, emergency preparedness
10. Failure to implement all those measures which fall between the emergency period following the damaging energy exchange and the final stabilization of the process	CSQ	Post-contact	Emergency preparedness

Table 3.2   (*Continued*)

Possible Control Failures	Relevant ECT Event (Typical)	ECT Level of Intervention & Causation	Relevant Hierarchy of Control
11. Failure to ensure relevant employees have sufficient awareness, knowledge, and/or skills	BE, IE, CE, CSQ	Prevention, pre-contact, post-contact	Administrative
12. Failure to provide adequate supervision	BE, IE, CE, CSQ	Prevention, pre-contact, post-contact	Administrative

BE — Breakdown event; IE — Intermediate event; CE — Contact event; CSQ — Consequence

useful inputs for RA teams selecting and evaluating controls. RA and RM will be elaborated on in Chapter 4.

3.6.5 Underlying factors

Underlying factors refer to organisational and managerial factors that are contributory to the failure of controls and occurrence of direct causes. They are similar to the latent conditions in the SCM, and basic causes and lack of control in the LCM. Underlying factors are often contributory in nature and their determination may have to depend on investigators' professional judgment, but identifying the underlying factors can lead to a broader and more sustainable impact on safety and health performance. Three types of underlying factors are identified in Figure 3.4: leadership, management system and culture.

In the context of WSH management, leadership refers to the attributes of senior managers that have an influence on

WSH. A senior manager is a person who has the ultimate authority and accountability, i.e. the buck stops with him or her. The leadership issues highlighted in Table 3.1 are possible factors that can contribute to the control failures and direct causes. Leadership issues will be discussed in more detail in the Chapter on Safety Culture.

A WSH management system refers to a set of policies, processes and structures that interact to assure or support WSH risk controls so that they maintain or improve the WSH performance of the organisation.

As an example, with reference to Table 3.1, a worker tripped over a wooden plank, i.e. breakdown event is loss of balance. The breakdown event was due to poor housekeeping (an unsafe condition) and the control failure is the failure to conduct frequent housekeeping in the worksite (i.e. failure to prevent the build-up of a hazardous "substance"). A relevant management system inadequacy could be inadequate processes for monitoring, measuring, analysing and evaluating the system's performance. More specifically, although the site might have defined housekeeping schedules, there was a lack of inspections to check that housekeeping was actually conducted. At the management system level, the focus is on the pattern of events and not on the specific incident. Thus, it is important to do a wider check on whether the control failure is related to more widespread inadequacies in the relevant management system elements. Management system inadequacies are usually inadequacies in the PDCA cycle.

Another important underlying factor is organisational culture. Safety culture has gained a lot of attention in the WSH literature over the years. Safety culture can be seen as a subset of organisational culture, and the latter is defined as, "a system of shared values [what is important] and beliefs [how things work] that interact with a company's people, organisational structures, and control systems to produce behavioural norms [the way we do things around here]" (Uttal, 1983). When applied to safety and health management, the "shared values", "[shared] beliefs" and "behavioural norms" refer to safety-related values, beliefs and behavioural norms (see Table 3.1).

Organisational culture is abstract and even people within the organisation may not be able to explain their own culture well, but they simply know that the culture is there. During the Safety Management in Context conference in 2013, Edgar Schein, an organisational culture guru, recommended that safety practitioners should focus on [management] processes and not 'safety culture'. He was also sceptical about the concept of safety culture, which he felt was an ill-defined concept. If we take Schein's comments at face-value, it seems to imply that the focus of any investigation of underlying factors should be on management processes. However, the author believes that Schein's comments were not meant to dilute the importance of organisational culture in incident prevention. His comments highlighted the importance of focusing on management systems so that shared values, beliefs and norms can be better managed. It is true that the concept of safety culture is abstract and its assessment is more subjective. Furthermore, culture constantly interacts with organisational structure processes, i.e., the organisation's management system. Therefore, a management system is a reflection of the culture and vice versa. Thus, instead of focusing on the abstract, from a WSH management perspective, it is more tangible to focus on the observable. It is recommended that when applying the ECT, if there is sufficient information to infer the shared beliefs, values and norms that contribute to an incident, the cultural problems and issues should be highlighted. Safety culture will be discussed in more detail in Chapter 7.

3.6.6 Implementing the Event Causation Technique

To utilise the ECT, a series of steps should be conducted iteratively during the analysis stage of the investigation. The steps in the application of ECT are as follows:

1. Describe the work context in which the incident occurred.
2. Create a timeline of the incident. The timeline should be as comprehensive as possible.
3. Select the key events (i.e. breakdown event, intermediate event(s), contact event and consequences) for detailed analysis.
4. For **each** of the key events in the incident sequence, a series of 'whys' are asked to facilitate the identification of direct causes, control failures and underlying factors. In accordance with the principle of multi-causation, within each component of the ECT, there can be more than one answer to the 'why'. The asking of 'why' questions should only stop when the answers reach underlying factors, i.e. leadership, management system, and/or culture. If necessary, the underlying factors can be inferred based on the evidence collected.
5. For communication and presentation purposes, it is necessary to simplify the why-analysis into the ECT diagram (see Figure 3.4 and Figure 3.7) to ensure that readers can understand the incident causation in one diagram. This involves compressing the why-analysis into more generic words and phrases (based on Table 3.1).

Table 3.1 is a useful prompt for applying the ECT and during the why-analysis and creation of the ECT diagram, the description of causes and factors should be based on the generic categories in the table. At the same time, each cause and factor should be clear enough to help readers understand what you are referring to. For example, if a relevant unsafe act is "failure to secure", the generic category should be followed a brief description such as, "did not secure dump truck tailgate".

3.6.7 Concrete hopper accident

The following case study, which we will call "Case Study 2", demonstrates how the ECT can be applied in an incident investigation.

3.6.7.1 Describe the work context

In Case Study 2, a worker ("worker A") was working inside a man-cage attached to a concrete hopper. The concrete hopper is a container used to carry wet concrete, and that has a release lever to allow workers to pour the concrete into formworks or other containers. In this case, the worker is needed to pull the lever to pour the concrete into a joint between two precast wall panels. The concrete hopper was lifted by a tower crane and the man-cage was connected to the concrete hopper. The work context can be summarised into the following statement, "Worker pouring concrete into precast joint while standing in man-cage lifted by tower crane and using concrete hopper". The statement points out the type of activity that was being conducted, and the equipment and plant that were used.

3.6.7.2 Create timeline

The timeline in Table 3.3 shows the events related to the case. The focus is typically on the events on the day of the accident, but relevant information such as the history of the hopper and man-cage, the start date of the project and other relevant activities, and the date when the operator and relevant workers first began working on site can also be included. Furthermore, the relevant evidence and other remarks should also be captured in the timeline succinctly. The timeline serves as a useful summary

Table 3.3   Timeline for Case Study 2

Date	Time	Event	Remarks
1 Dec 2018	—	Project started	Project supposed to end in Dec 2021; contract between developer and contractor
13 Mar 2019	—	Worker A started working on site; induction training conducted.	Inducted by WSH Officer (WSHO); induction record and WSHO statement
1 May 2019	—	Hopper first installed with man-cage; crane operator started working on site	Inducted by WSH; see induction record and WSHO statement
May–Sep 2019	—	Man-cage used on several joint castings	Crane operator and lifting supervisor statements
2 September 2019	0830	Worker A starts work in man-cage; crane operator lifts man-cage	Co-workers and lifting supervisor statements
	0832	Hopper was lifted to the 6–7th level joint	Co-workers and lifting supervisor statements
	0835	Crane operator needs to adjust the height of the hopper to facilitate concreting	Co-workers and lifting supervisor statements
	0836	Hopper became dislodged suddenly	Co-workers and lifting supervisor statements
	0840	Site supervisor contacted WSHO	Site supervisor and WSHO statement
	0841	WSHO called ambulance	Site supervisor and WSHO statement
	0850	Worker A sent to hospital	Site supervisor and WSHO statement

of the key information for the incident, which is very useful for guiding the investigation and analysis phases.

3.6.7.3 Describe incident sequence

The incident sequence is created based on the timeline. The events that are uncertain or were based on the opinions of investigators need to be highlighted using a symbol or a different colour. These uncertainties can then be further validated using new evidence, which become "to-dos" for the investigators. The incident sequence shown in Figure 3.5 is the finalised incident sequence. Investigators will face situations where there are several competing hypotheses about what happened during the incident and several possible incident sequences can be drawn. Investigators need to use evidence to help them eliminate unlikely incident sequences and decide on the most likely incident sequence.

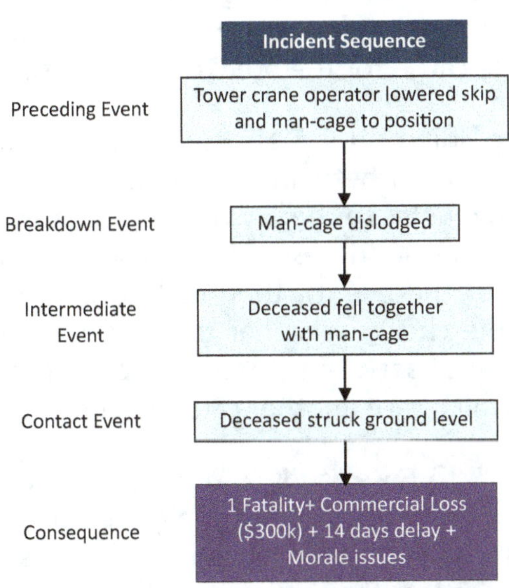

Figure 3.5   Case Study 2 incident sequence

3.6.7.4 Conduct why-analysis

Once the incident sequence is established, the why-analysis should be conducted on each incident event. In Case Study 2, the discussion is focused on the breakdown event because during investigation it was assessed that it was unlikely for the intermediate event and contact event to be prevented once the breakdown event is initiated. Furthermore, during the investigation it was established that the emergency response was conducted promptly and in accordance to the emergency response plan. Sometimes, investigators can choose to focus on some of the key events due to time constraints, lack of resources or unavailability of evidence, but **it is emphasised that each incident event should be analysed as much as possible**. With reference to Figure 3.6, the qualitative analysis process is based on one simple question, "why did the event or cause occur?" For instance, "why did the man-cage dislodge from the concrete hopper?" In this case, the answer is "the man-cage dislodged because when the crane operator was lowering the hopper and man-cage into position (preceding event), the man-cage unhooked from the hopper (the cause for the breakdown event)."

Two basic guiding principles need to be consulted for each cause, the first being whether the cause is "necessary". The first guiding principle helps to check whether the cause or factor is a true cause. If it is not "necessary" (i.e. if it does not influence the existence of the event), then it should be removed. The second principle checks whether the cause is "sufficient". A cause may be necessary, but it may not be sufficient to cause the event, in which case another cause needs to be included.

Coming back to Case Study 2, during the activity of lowering the hopper, the unhooking of the man-cage is necessary to cause the man-cage to be dislodged. This means that without the unhooking action, the man-cage will not be dislodged. The unhooking is also sufficient to cause the dislodgement, i.e. other causes like strong wind or high temperature do not need

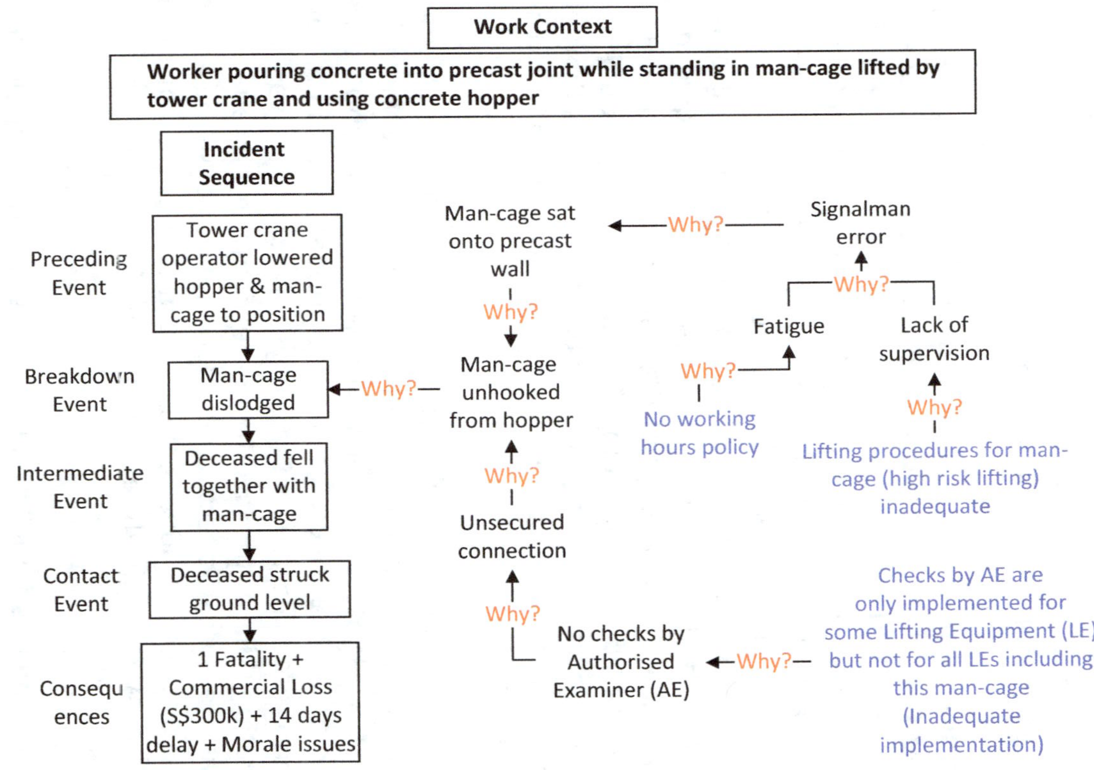

Figure 3.6   Why-analysis for the breakdown event

to be present to result in the man-cage dislodging. The interrogation continues for "man-cage unhooked from hopper". In this case, two causes were identified. Firstly, in the midst of the lowering process, the man-cage sat onto the precast wall, which produced a vertical reaction force, contributing to the man-cage becoming unhooked. Thus, "Man-cage sat onto precast wall" is a necessary cause. However, for the man-cage to 'complete' the unhooking action, pushing the man-cage against the precast wall alone is not sufficient to produce the "unhooking". The connection between the man-cage and hopper must have been unsecured, which allows a vertical force to push the man-cage out of the hopper. Thus, "unsecured connection" is included as another cause. The same process continues for each cause, guiding the investigator in his or her evidence-gathering process. The why-analysis can be guided by the taxonomy in Table 3.1, but the investigators should not be constrained by the taxonomy. The why-analysis should end when the investigator uncovers the underlying factors that can produce fundamental systemic improvement to the organisation.

Even though, many why-analyses are conducted individually, it is best done in a team with a whiteboard or a wall with space for plenty of post-its. The why-analysis should be reviewed as the investigation progresses and when new evidence are collected. If the analysis is done in a team, the facilitation skills of the investigator leading the why-analysis is critical to ensure its effectiveness. During the analysis, be it with a team or done individually, the investigator(s) must remember the following guidelines (acronym of GESIS) during the analysis:

1. Be **Generous** initially and do not strike out any possible events, causes or factors
2. Be **Evidence**-driven and highlight hypotheses or opinions

3. Be **Specific** when responding to the "whys"
4. Be **Inclusive** and prompt different team members or other colleagues to provide inputs or critique the analysis
5. Be **Structured** as the team or investigator works through one event, cause or factor at a time and list out actions or to-dos along the way

3.6.7.5 Summarise using the Event Causation Technique diagram

The why-analysis can extend into several pages and can be confusing to the readers, especially personnel not directly involved in the accident. Thus, the why-analysis is summarised into a simpler ECT diagram (Figure 3.7). The ECT diagram should also include the summary of the analysis for the contact event and

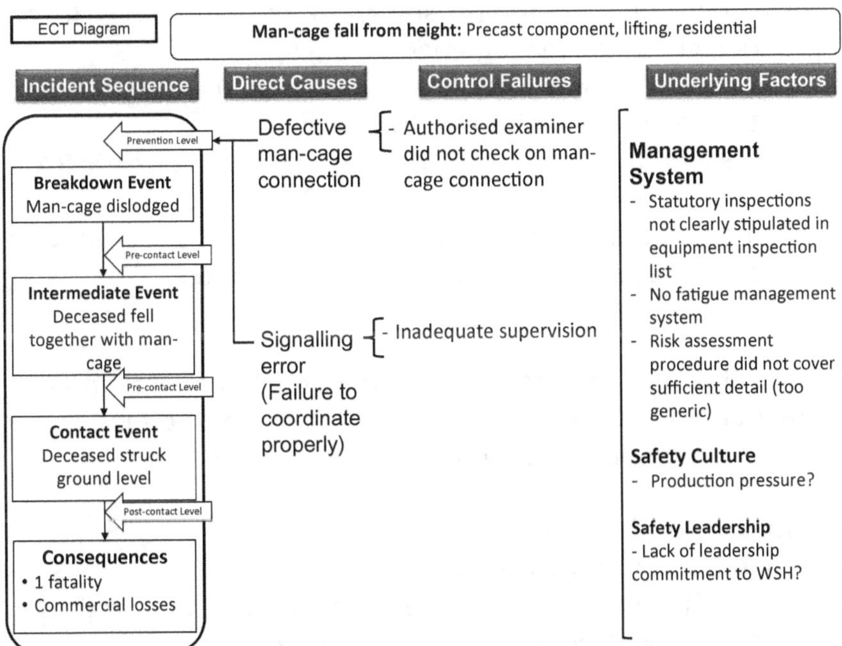

Figure 3.7   ECT diagram for Case Study 2

consequences. This is when each column in the ECT diagram, incident sequence, direct causes, control failures and underlying factors, are filled in based on the why-analysis. Examples of complete ECT diagrams can be found in Goh and Soon (2014).

Using the ECT diagram, recommendations can then be made to improve the WSH performance of the organisation. Each recommendation should be linked to different parts of the ECT diagram. Based on the ECT diagram for Case Study 2 (Figure 3.7), some possible recommendations are:

1. Top management to impose an operational requirement for engineers and supervisors to ensure that statutory equipment are checked before they are used.
2. Human resource to develop a set of work hour policies and a fatigue management system that is applicable to all site personnel (including sub-contractors); policies to be endorsed by top management.
3. To review lifting procedures and RA to differentiate between low- and high-risk lifting operations.
4. To review RA procedures to determine when generic RA is allowed.
5. To ban all man-cage usage until RA for lifting is reviewed; sites to use scaffold as substitution in the interim.

The clear linkage provides a stronger rationale for the recommendation. The recommendations can then be monitored using an action plan that clearly answer the following:

- Who to do what and by when?
- When is the next review date?
- Who is the management representative to ensure progress?
- How to track effectiveness of measures?

More examples of ECT diagrams can be found in Chapter 10.

3.7 Systems Archetypes

Even when systemic models and investigation methods uncover underlying factors (leadership, management system and culture) that contributed to the incident, the question of why these underlying factors arose still needs to be answered by management. As highlighted in Chapter 2, systems thinking literature had accumulated a set of systems archetypes which are useful for identifying fundamental organisational dynamics that could lead to poor performance, including poor WSH performance. Chapter 2 highlighted one of the archetypes, "shifting the burden". Table 3.4 summarised ten archetypes and provided WSH examples and management guidelines related to the archetypes.

These archetypes are essentially common organisational and management issues found in a wide range of contexts. By being aware of the systems thinking archetypes, the organisation's management can identify possible leverage points they can focus on so as to improve WSH and other aspects of the organisation. Table 3.4 serves as a good primer for understanding deep rooted dynamics that contribute to accidents and ill-health in workplaces.

3.8 Summary

Incident investigation is an important part of WSH management. When an incident happens, there must be deliberate attempts to identify the patterns in the way things were done so that these patterns can be changed to prevent a recurrence of the incident. These patterns occur because of some underlying factors such as the way the management system is constructed and the characteristics of the safety culture of the organisation. Thinking of how the system influenced the occurrence of an event is a key concept in systems thinking and it is an important skill for any manager.

Table 3.4   Systems archetypes

No.	Systems Archetype	Description	WSH Examples	Management Guidelines
1	Balancing a process with delay	When working towards a goal, a party receives delayed feedback about the outcome of their effort. The party is not conscious of or does not have full knowledge of the delay. This leads to inappropriate actions (too much or too little) to close the gap between current performance and the goal. The party may also give up even before any significant results.	"We have not seen any improvement since implementation of the new WSH programme. Should we continue with it? Perhaps we should just cancel it."	• When feedback is delayed, be patient. • Build information system that provides feedback quicker. • Actively seek the inputs of frontline staff.
2	Eroding or drifting goals	The two ways to close the gap between current performance and a goal is either to improve the performance or to reduce the goal. However, it takes time and resources to improve performance, while it takes minimal effort to reduce the	"It is fine to reduce the goal of having safety inspections everyday to once a month, as we need to get the work done by this week. We will remind the workers to be more careful."	• Hold the vision, e.g. leaders need to reiterate the importance of WSH goals despite the difficulties on the ground • Link goal setting mechanisms to parties not within the system, so

		goal. This archetype describes how individuals or organisations take the easy way out of reducing the goal so that the pressure to improve is removed, but the actual performance is eroded over time.	"Even though our goal is to review 4 RAs every month, we simply don't have the time. Let's review 2 this month and ask management to reduce the goal."	that it is harder to erode the goal, e.g. make WSH commitments and goals public and accountable to clients or unions.
3	Fixes that fail	There is a problem arising and it is usually very urgent. There is a quick solution to the problem, but it has negative consequences that will aggravate the problem.	"Our project team doesn't have the competency to conduct RA. Let's employ consultants to conduct the RA for us." "Everyone is afraid of the new virus and are clamouring for face masks. We should just give out masks to allay their fears. We'll deal with the shortage of face masks later."	• Use stopgap measures sparingly to "buy time" • Focus on long term solutions • Communicate the long-term consequences of fixes that fail

(Continued)

Table 3.4   (*Continued*)

No.	Systems Archetype	Description	WSH Examples	Management Guidelines
			"There are not enough technicians to implement the maintenance programme. Let's reduce the maintenance requirements so that it is quicker to finish each maintenance job. It might mean more breakdowns in the longer run, but let's deal with the workload issue first."	
4	Shifting the burden (addiction)	As an extension of fixes that fail and eroding goal, there are two options to solve a problem, either by using a fundamental solution that is more difficult to implement or use a short term	"It is too difficult to convince workers of the importance of WSH. Just punish those who flout the WSH rules even if it means that it	• Use stopgap measures sparingly to "buy time" • Focus on long term or fundamental solutions • Communicate the long-term consequences of

		fix that has side effects on the capability to execute the fundamental solution. Organisations that get addicted to the short-term solution becomes unable to execute the fundamental solution in the longer term.	will become harder to communicate with them in the future."	becoming "addicted" to the short-term solution
5	Limits to growth	A party is experiencing growth and the growth seems to be self-sustaining. At some point, the growth slows, a halt or even reverse into decay due to a "limit" or constraint in the system, e.g. resistance from different parties, resource constraint, lack of knowledge and insufficient time.	"The behaviour based safety programme was growing in terms of participation among the different project teams. However, the programme came to a sudden halt as some of the managers who were not involved perceive the programme as too expensive."	• Identify and focus on removing the limits early • Do not push on the growth

(Continued)

Table 3.4   (*Continued*)

No.	Systems Archetype	Description	WSH Examples	Management Guidelines
6	Growth but under-investment	A special case of limits to growth. Growth can be sustained if a party invests in its capacity to faciliate further growth. However, the investment to build capacity is costly and there is pressure to under-invest to conserve resources.	"Our customers are demanding us to do more in terms of WSH management, but that requires more investment in WSH training and equipment. Let's wait and see if the demand from customers is just a passing fad."	• Maintain continuous investment in WSH management capacity • Be aware of the WSH management norms and stay ahead of the trends
7	Escalation	When two parties are engaged in a "zero-sum" game where the gain of one party is perceived as the loss of the other party. In this situation, unhealthy competition can easily arise, and a vicious cycle can be created where the	"The market is so competitive on price that everyone cuts their safety budget so much that safety performance is affected."	• Identify the indicator (e.g. price, sales figures and WSH budget) that triggers the comparison and competition • Change the indicator to promote a virtuous cycle or healthy competition

		aggressive actions by one party triggers further aggression from the other party. If used wisely, the vicious cycle can be converted into a virtuous cycle, where the competition is used as an external motivation for improvement. However, reliance on external motivation for improvement can expose an organisation to possible vicious cycle that leads to poorer performance.	"The other department is putting in a lot of effort into WSH; we should do the same if not we will look bad." "The other department is not doing anything on WSH and using all their resources to increase sales. We look like fools spending so much effort on WSH. We better do the same."	(e.g. compete on number of hazards reported with acceptable level of quality) • Promote larger goals that emphasise collaboration (e.g. collaboration at industry level to develop tendering practices that are sustainable)
8	Success to the successful	Two parties compete for limited resources and the party that is more successful gets more resources leading to more success. The other party gets less resources leading to less success.	"Unlike the other manager, this manager is doing a great job in implementing the WSH measures. Let's give him a few more WSH coordinators to help him with his interventions."	• Bears some similarity with Escalation • Promote common goals and teamwork so that the successes are shared • Stipulate different goals for different teams so that the definition of success can be differentiated and de-coupled.

(Continued)

Table 3.4 (*Continued*)

No.	Systems Archetype	Description	WSH Examples	Management Guidelines
9	Accidental adversaries	Two parties cooperate to increase the chances of success but ended up limiting each other's success. The degeneration starts when one party takes action that has unintended negative consequences on the other party, which leads to retaliations and hence a vicious cycle.	"We are supposed to cooperate with company B on WSH initiatives so that our clients can see a consistent approach across the whole supply chain, but whenever we meet out clients, they always brag about their WSH initiatives, which make us look bad. We have to do something about this."	• Ensure open communication between the partners • Evaluate possible impact on partnership prior to any key actions • Make the benefits of the collaboration explicit to each party
10	Tragedy of all common goods (tragedy of the commons)	Two or more parties share a common resource that is limited. However, each party use the resource from their own perspective without considering that the resource is limited. The	"The centralised WSH team is being overloaded because every department is relying on them to conduct risk	• Manage the shared resources by making the use of the resources explicit and ration the resources • Make each party self-regulate

		common resource can end up depleted and all parties are impacted.	assessment, safety inspection and other WSH management tasks. Now their support is so weak that it's like they are non-existent." "Everyone assumes that the world has unlimited supply of fish, clean air and water and other natural resources. Everyone keeps polluting the air, cutting down trees and digging up natural resources, and that is why we have climate change."	and become more independent of the shared resource • Educate all parties of the limitations of shared resource

The investigation process is a qualitative, systematic and evidence-based process. The investigator needs to connect the investigation, a reactive process, with proactive RA so that relevant risk controls identified in the RAs are reviewed and future RAs are strengthened. The ECT is a simple and flexible method for incident analyses. Despite its simplicity, it is meant to be thorough, and the final ECT diagram should provide a useful summary of an incident.

Even when the underlying factors are identified, the management will need to understand the system dynamics that led to the underlying management system, leadership or cultural problems. One way to help the management understand the system dynamics is to evaluate the ten systems thinking archetypes highlighted in this chapter. The systems archetypes describe common management issues and provide some guidelines for resolving those systems thinking or system dynamics issues.

Review Questions

1. "An accident is a matter of luck." Discuss the validity of this statement and how management should see the role of luck in accidents.
2. Explain the difference between corrective actions, preventive actions and actions for continual improvement.
3. Explain how incident investigation and RA are related.
4. Explain what it means for investigators to understand the "retrospective" nature of incident investigation.
5. Provide examples of the four types of evidence in an incident investigation.
6. Practise the ECT method on an accident case that you find on the internet.
7. Compare and contrast the different systems thinking archetypes. Which of them apply to your own situation?

References

Bird, F. E., Germain, G. L., and Clark, M. D. (2003). *Practical loss control leadership*. Duluth, Georgia: Det Norske Veritas (U.S.A.), Inc.

British Standards Institute (2007). "BS OHSAS 18001:2007 Occupational health and safety management systems — Requirements." BSI, London.

British Standards Institution (2004). "BS EN ISO 14001:2004 Environmental management systems — Requirements with guidance for use." British Standards Institution, London.

Chua, D. K. H., and Goh, Y. M. (2004). "Incident causation model for improving feedback of safety knowledge." *J. Constr. Eng. and Manage. — Am. Soc. of Civ. Eng.*, 130(4), 542–551.

Dekker, S. (2017). *The field guide to understanding 'human error'*. Florida: CRC Press.

Energy Institute (2008). "Guidance on investigating and analysing human and organisational factors aspects of incidents and accidents." <http://www. energyinstpubs.org.uk/ tfiles/1368180505/817.pdf>. (accessed May 10, 2013).

Goh, Y. M., and Soon, W. T. (2014). *Safety Management Lessons from Major Accident Inquiries*, Pearson, Singapore.

Haddon Jr, W. (1973). "Energy damage and the ten countermeasure strategies." *Human Factors*, 15(4), 355–366.

Health and Safety Executive. (2004). *HSG245 Investigating accidents and incidents — A workbook for employers, unions, safety representatives and safety professionals*. London: HSE.

Kim, D. H., and Lannon, C. P. (1997). Applying systems archetypes. California: Pegasus Communications Inc.

Lappalainen, J., and Perttula, P. (2017). "Accident investigation techniques". https://oshwiki.eu/wiki/Accident_investigation_techniques#cite_note-13. (accessed September 1, 2020).

Uttal, B. (1983). "The corporate culture vultures." *Fortune*, Oct. 17, 66–72.

WorkCover Queensland. (n.d.). Incident Investigation Form. Retrieved from https://www.worksafe.qld.gov.au/__data/assets/word_doc/0019/131761/incident-investigation-form.doc

Workplace Safety and Health Council. (2013). *Workplace Safety and Health Guidelines — Investigating Workplace Incidents for SMEs*. Singapore: WSHC.

CHAPTER 4

Workplace Safety and Health Risk Management

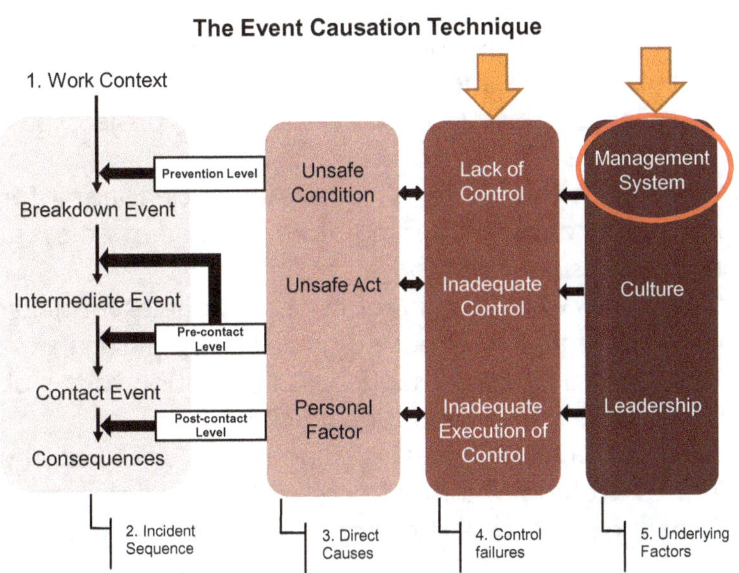

Risk management (RM) is an important aspect of the management system as it forms the planning part of the workplace safety and health (WSH) management framework (see Event Causation Technique (ECT) diagram above). The end products of an RM process are the controls to prevent incidents and reduce the severity of injuries and ill health.

4.1 Introduction

Risk management (RM) is not unique to workplace safety and health (WSH) management. According to the ISO standard, BS ISO 31000:2018

Risk Management—Guidelines (British Standards Institute, 2018), risk is the effect of uncertainty on objectives. The following definitions and comments are adapted from BS ISO 31000:2018 and ISO Guide 73:2009 (British Standards Institute, 2009):

- An effect is a positive and/or negative deviation from the expected.
- Risk can address, create or result in opportunities and/or threats.
- Objectives can be in relation to different aspects of an organisation such as financial, productivity, WSH, and environment. At the same time, objectives can be at different levels of an organisation; e.g., the strategic, project, product and process levels.
- Uncertainty is the partial or complete deficiency of information related to understanding or knowledge of an event, its consequences or likelihood.
- Risk is usually expressed in terms of risk sources (or hazards), potential events (occurrence or change of a particular set of circumstances), their consequences (in relation to the objectives) and their likelihood (or chance).
- RM is a systematic process of identifying, analysing, evaluating, controlling and monitoring risk. It is implemented to help assure achievement of organisational objectives.

In the context of this book, the objectives we are concerned with are WSH objectives. When organisations assess WSH risk, they typically refer to negative effects such as accidents and pollutions, but there are also potential positive effects, like higher employee morale, higher productivity, improved reputation, and new business opportunities. However, these positive effects are usually only apparent at organisational or strategic levels. Most of the WSH RM is conducted at the operational level and focuses on operational issues. We will first discuss the importance of RM in WSH and the role that it plays. We will then cover the RM process and key WSH hazards and controls. Finally, we will discuss some of the problems and difficulties with WSH RM.

4.2 Workplace Safety and Health Risk Management and Assessment

RM, which includes risk assessment (RA), is a cornerstone for the performance-based approach to the regulation of WSH. The term RM is sometimes used interchangeably with RA, but there are differences that will become obvious when we discuss the steps in RM subsequently.

4.2.1 Background

One of the key documents that resulted in the common use of risk assessment in WSH is the UK Roben's report in 1972 (Committee on Safety and Health at Work, 1972), which highlighted the following:

- Health, safety and welfare at work could not be ensured by an ever-expanding body of legal regulations enforced by an ever-increasing army of inspectors;
- The primary responsibility for ensuring health and safety should lie with those who create risks and those who work with them; and
- The law should provide a statement of principles and definitions of duties of general application, with regulations setting more specific goals and standards.

Singapore adopted a similar approach in 2006 when the WSH Act was enacted. At the time of the enactment, then Minister of Manpower Dr Ng Eng Hen highlighted the following principles:

- Reduce risks at source
- Promote industry ownership of standards and outcomes
- Penalise poor management (ensure management accountability)

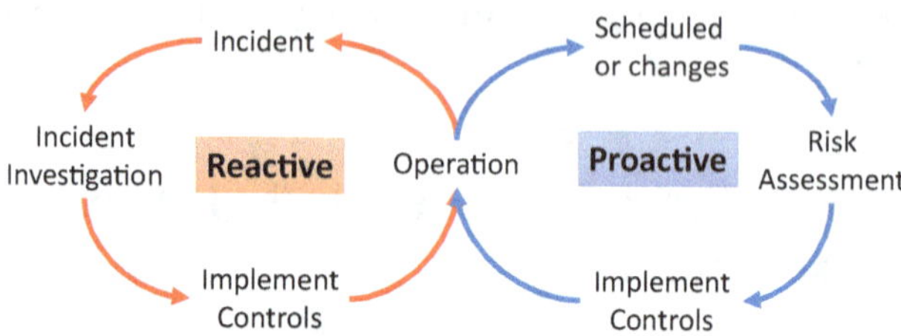

Figure 4.1   Risk assessment (RA) allows stakeholders to be proactive

Thus, for the industry to reduce risk at source and own standards and outcomes, employers and duty holders are expected to conduct RA (as part of their RM) to proactively reduce the WSH risk of its operations and activities (see Figure 4.1). Even though incident investigation is an important aspect of WSH management, it is reactive and dependent on the occurrence of incidents which could have resulted in severe injuries and losses. Thus, RA, instead of incident investigation, should always be the key focus of WSH management. RAs must be scheduled prior to all operations so that controls can be implemented. RAs should also be conducted whenever there are changes in the operations, after significant incidents, and when organisations receive additional or new WSH information that shed light on hazards, possible unintended events, and likelihood and potential consequences.

4.2.2 Risk assessment and incident causation models

The RA process is very much related to incident causation models. This is because incident causation models tell us why and how incidents happen and RA aims to identify possible incidents, their causes and corresponding controls and systems to manage the likelihood and/or severity of the incidents (see Figure 4.2). During RA, it is important to identify the controls to manage the basic causes and immediate causes (hazards) to prevent incidents from happening.

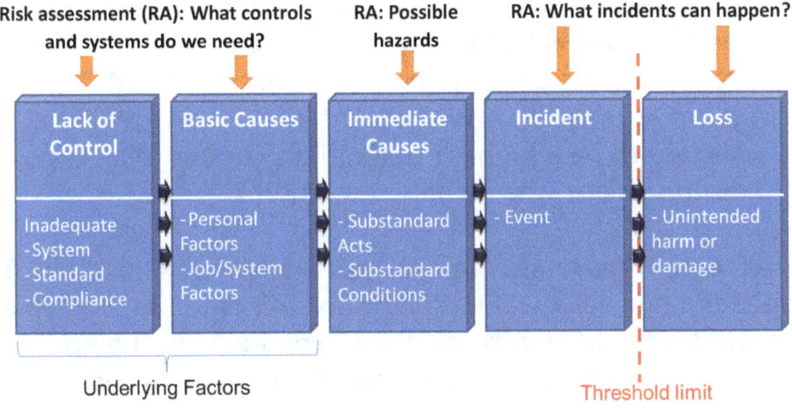

Figure 4.2   Risk assessment and incident causation model (LCM)

The possible controls can be guided by Haddon's Energy Transfer Model (ETM) (see Chapter 2) and the hierarchy of control, which will be discussed later.

4.2.3 Conceptual basis of risk assessment

The UK Health and Safety Executive (HSE) published the seminal document, "Reducing Risk, Protecting People" in 2001. It provides a useful conceptual basis for WSH RA. Some key concepts presented include the following:

- Risk is a given. There are hazards everywhere, even if we do not do anything. Thus, it is theoretically impossible to have zero risk.
- Numerous factors influence risk perception of man-made hazards. For example:

  - if a hazard poses a risk to things that are important to us, we will perceive the hazard to be a higher risk.
  - if we do not understand a process giving rise to the hazard, i.e. we have less knowledge and competence, we will perceive the hazard to be a higher risk.
  - if the risk is not equitably distributed, we will perceive the hazard to be a higher risk ("are we the only ones that are exposed to the risks?")

- o   if we cannot control our exposure to the hazard, we will perceive the hazard to be higher risk ("do we have the resources and capabilities to control the hazards?")
- o   if we assumed the risk involuntarily, we will perceive the hazard to be a higher risk ("did we choose to be exposed to this risk or was it imposed on us?")

As seen, RA is heavily influenced by human minds, cultures, judgement and values. Thus, it is difficult to establish the "objective and true risk", which may not even exist. In the end, it is about the tolerability and acceptability of risks by individuals, organisations and society. In addition, a major factor is the amount of resources (e.g. time, personnel, equipment, and money) that should be invested to control the hazards, as compared to the aggregated consequences suffered by society, community, organisations and groups, when accident(s) and/or ill health occur(s). Thus, RA is essentially a cost–benefit analysis, which balances the costs of controlling risk and the benefits of having the risk controlled. However, it must be noted that organisations and individuals are expected to take their RA seriously and they should always be proactive, conservative and comprehensive during RA. This means that the cost should not become an excuse to not implement risk controls.

As highlighted in Chapter 1, a key guiding principle of RA is "as low as reasonably practicable" (ALARP). The risk levels of all hazards should be ALARP, which means that the costs involved in reducing the risk level further would be grossly disproportionate to the benefits gained. With reference to Figure 4.3, the risk level of a hazard can be placed on a scale with *Broadly acceptable*, *Tolerable* and *Unacceptable* regions. The *Unacceptable* region is where the risk level of a hazard is too high and no person should be exposed to this risk regardless of the benefits gained. On the other hand, the *Broadly Acceptable* region indicates that the risk level is widely regarded by society to be acceptable because the inherent risk is low, or the controls are very effective. The *Tolerable* region is where the risk level is significant, but tolerable such that the benefits justify the exposure

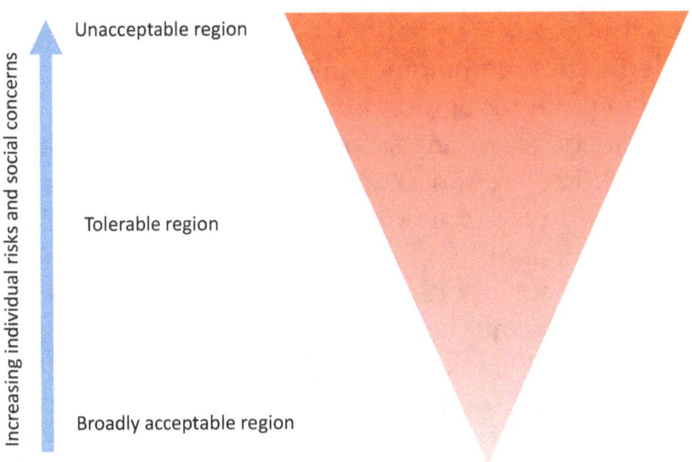

Figure 4.3   HSE Framework for Tolerability of Risk; adapted from http://www.hse. gov.uk/risk/theory/r2p2.pdf (Pg. 42)

to the hazard. Some examples of benefits include employment, lower cost of production, personal convenience, and continued supply of basic needs like energy, food and water. In each of these regions, the risk level of the hazard should be kept ALARP. The concept of ALARP is especially important in the *Tolerable* region. For a hazard to be tolerated, it must be properly risk-assessed, the residual risk level must be ALARP, and the risk level must be periodically reviewed to ensure that it is ALARP.

4.3 Risk Management Regulations

As an example of an RM legislation or framework, we will discuss the Singapore WSH (Risk Management) Regulations 2007 (RM Regulations). In Singapore, all employers, self-employed and principals[1] who engage contractors, are required to

---

[1] In accordance to the Workplace Safety and Health Act, "principal" means a person who, in connection with any trade, business, profession or undertaking carried on by him, engages any other person otherwise than under a contract of service (i.e. an employee) — to supply any labour for gain or reward or to do any work for gain or reward. Typically, a principal refers to the client in a client–contractor relationship.

comply with the RM Regulations. The RM Regulations is a very concise legislation containing only eight regulations, and is a very fundamental regulation in the WSH regime. Anyone that employs or engages others for work has to conduct RA (see RM Regulations). RA is defined as "the process of evaluating the probability and consequences of injury or illness arising from exposure to an identified hazard, and determining the appropriate measures for risk control". A "hazard" is defined as anything with the potential to cause bodily injury, and includes any physical, chemical, biological, mechanical, electrical or ergonomic hazard. According to the RM Regulations, "risk" refers to the likelihood that a hazard will cause a specific bodily injury to any person. This definition is more specific than that provided by ISO 31000 because it is contextualised to WSH.

Employers, self-employed and principals are required to "eliminate foreseeable risk", but if it is not "reasonably practicable" to eliminate the risk, employer, self-employed person or principal are to implement "reasonably practicable measures to minimise the risk" and safe work procedures to control the risk. The control measures identified in the RM Regulations can be found in Figure 4.4, which is also known as the hierarchy of control. The control measures are defined as follows:

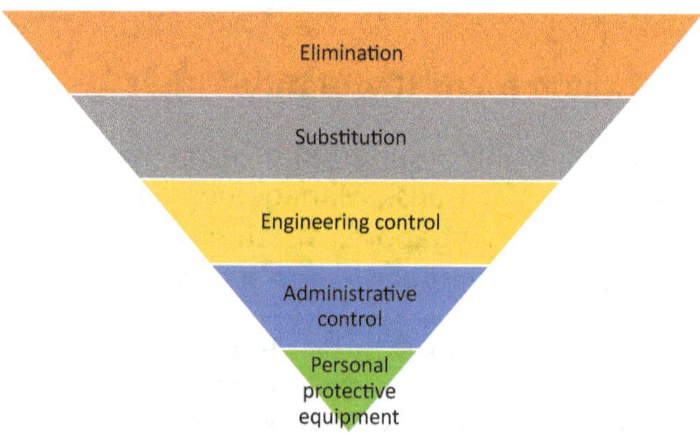

Figure 4.4   Hierarchy of control

- "substitution" means the replacement of any hazardous material, process, operation, equipment or device with less hazardous ones.
- "engineering control" means the application of any scientific principle for the control of any workplace hazard; and includes the application of physical means or measures to any work process, equipment or the work environment, such as the installation of any barrier, enclosure, guarding, interlock or ventilation system.
- "administrative control" means the implementation of any administrative requirement which includes a permit-to-work (PTW) system.
- "personal protective equipment (PPE)" refers to equipment, such as helmets, safety boots, harnesses and safety goggles, that a person wears to protect against hazards.

As noted earlier, Haddon's ETM is also a useful guide for the identification of possible risk controls.

The hierarchy of control indicates that elimination is the most effective approach, while PPE is the least effective. This is because PPE only minimises the impact of hazards and is heavily reliant on the user using the PPE correctly at all times. In contrast, elimination removes the hazard and the need for any form of control. Substitution and engineering control are highly desirable, but many-a-times, organisations rely heavily on administrative controls and PPE to manage WSH risk because they are more easily implemented. It should be noted that the control measures should be implemented concurrently to reduce the consequences if any of the control measures fail unexpectedly. For example, a worker working at height should be protected by barricades, but if the worker needs to reach over the barricade during the course of his/her work, then a fall protection system (full body harness, lanyard with personal energy absorber and anchorage), which is a form of PPE, should be used by the worker to reduce the risk of falling from height. Table 4.1 shows some examples for each type of control in the hierarchy of control.

Table 4.1   Examples of controls

Type of Control	Example
Elimination	<ul><li>Using long arm roller paint brush instead of working on scaffolding</li><li>Designing windows that can be cleaned from indoors (e.g. can be turned inwards) to eliminate the need for workers to clean the windows from outside and at heights</li></ul>
Substitution	<ul><li>Substituting flammable and potentially toxic solvent- based paints (i.e. may release flammable Volatile Organic Compounds) with water-based (also known as acrylic emulsions) paints</li><li>Substituting a noisy machine (>85 dBA) with a less noisy machine (~60 dBA)</li></ul>
Engineering control	<ul><li>Barricades along open edge</li><li>Ventilation to remove toxic chemicals in the atmosphere</li><li>Sensors to detect human motion in the vicinity of a machine and stopping the machine when the sensors are triggered</li></ul>
Administrative control	<ul><li>Work at heights training for workers working in mobile elevated work platforms</li><li>Permit to work for workers entering confined space</li><li>Warning signs at the bottom of a scaffold under erection to remind workers of the need to stay away</li></ul>
Personal protective equipment (PPE)	<ul><li>Use of ear plugs by workers working in noisy area(s)</li><li>Use of eye protection for workers using cutting machines</li><li>Dust mask for workers cleaning dusty areas</li></ul>

ISO 31000:2018 uses the term risk treatment instead of risk control. Section 6.5.2 of ISO 31000:2018 highlights that risk treatment options are not necessarily mutually exclusive and can include the options stated in Table 4.2.

A "safe work procedure" (SWP) is frequently seen as a type of administrative control measure but the Technical Advisory for Working at Height (Workplace Safety and Health Council, 2008) describes SWP as, "a set of systematic instructions on how work can be carried out safely. Arising from the risk assessment, a set of SWP should be written for various jobs on site. The SWP provides a step by step account of how jobs are to be executed, who is in charge of these jobs, what safety precautions must be taken (based on the risk assessment made earlier) and what kind of training is necessary for the workers doing these jobs. The PTW system has to be integrated with the SWP so that the supervisors are made aware of the safety requirements and checks. The SWP must be communicated to everyone involved in the job so that each is aware of the role they play in it."

Therefore, an SWP is not a control measure, but a documentation of different control measures integrated with the work steps. Thus, for it to be effective, it must be written in a very detailed fashion and must include the risk controls to be taken in the course of the work and during an emergency. In addition, SWPs must highlight suitable PPE that must be provided to the persons carrying out the work. Thus, the SWP is essentially a set of work instructions that includes risk controls identified in the RA. There is no fixed format for SWPs, but they should be written such that they are easily comprehended by the people conducting the work—like how assembly instructions for self-assembled furniture include step-by-step guidance on how to put the furniture together as well as safety measures, such as providing a mat to protect the person assembling the furniture from injuries such as knee injuries due to prolonged kneeling on a hard floor.

Table 4.2   Risk treatment options described in ISO 31000:2018

ISO 31000:2018	Comments
Avoiding the risk by deciding not to start or continue with the activity that gives rise to the risk	Similar to elimination in the hierarchy of control, but instead of eliminating the hazard, the focus is on stopping work so as to prevent exposure to the hazard
Taking or increasing the risk in order to pursue an opportunity	ISO 45001:2018 encourages organisations to proactively look for opportunities to improve WSH performance. However, due to the negative nature of accidents and ill health, most WSH RA do not focus on opportunities.
Removing the risk source	This is the same as elimination of hazard
Changing the likelihood	Most WSH risk controls change likelihood rather than consequences.
Changing the consequences	Reducing consequences of an incident can be achieved through reducing the amount of energy or substances stored, and improving the resistance and resilience of individuals or organisations.
Sharing the risk (e.g. through contracts, buying insurance)	Also known as risk transfer, where specialist contractor or organisations and individuals with suitable competencies are selected to share the risk. When contractors are engaged, the principal who engaged the contractors are still expected to conduct RA.
Retaining the risk by informed decision	This option is only viable if the risk is already ALARP. Organisations must be able to demonstrate ALARP and that the decision was an "informed decision".

RA is expected to be recorded, disseminated and reviewed. The documentation of RA is necessary because it helps to ensure a detailed analysis of the work and establishes account-ability, but there are many challenges in actual implementa-tion, which will be discussed subsequently. RA should be reviewed every three years, when there is an incident, when there is significant change to the work and when relevant WSH information becomes available.

4.4 Overview of Risk Management Process

The Code of Practice on Workplace Safety and Health (WSH) Risk Management (Workplace Safety and Health Council, 2015) describes RM processes that an organisation should implement (see Figure 4.5). As can be seen, RA is a critical step of RM.

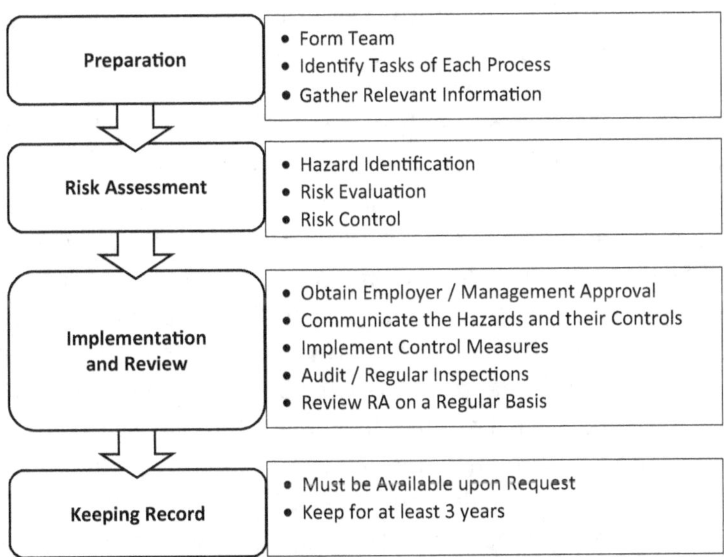

Figure 4.5 Risk management process (Workplace Safety and Health Council, 2015)

4.4.1 Preparation

The RM process starts with the formation of an RM team, or RA team if the workplace is relatively small and have less variety of activities and hazards. An RM team consists of an RM leader (or champion) and multi-disciplinary team members who have knowledge of the different hazards and relevant controls. The RM leader or champion must have attended suitable RM training and be able to formulate an RM plan customised to the needs of the organisation. An RM plan is essentially a plan to implement the different controls so as to manage the WSH risk level of the workplace. Another important responsibility of the RM leader is to select the relevant team members with the right RM training, relevant expertise and suitable job responsibilities. The RM leader should have the ability to chair RA meetings effectively, i.e. he/she should have the ability to facilitate discussions, get members to contribute different opinions and help the group arrive at agreed decisions efficiently. The RM leader also needs to ensure proper documentation and record retention. Some possible team members include management staff, site engineers, technicians, safety personnel, supervisors, operators, contractors, and suppliers. The roles and responsibilities of the team members must be clearly established and only personnel trained in RA can be appointed as team leaders. For large organisations and organisations with a wide range of activities, there will be a need for RA teams to be created. Each RA team will be focused on specific activities or areas in the workplace and they will report to the RM team. The RA team leader will also be a member of the RM team. If the hazards and controls need specialist knowledge, or if the workplace does not have people versatile with RM/RA, it may be useful to engage consultants (e.g. engineers, and technical specialists) to assist with the process. However, the consultants should be there to assist and the RM/RA teams must still conduct the RM process and make the key decisions such as determination of risk levels and selection of controls.

The RM team will have to create an inventory of work activities and scope the RA. Scoping is essentially determining the boundaries for each RA. This can be based on the activities, location, type of personnel, material, etc. The scope of RAs will also be guided by the organisation's policy, objectives, targets and programmes. For example, a developer may want to focus on hand and finger injuries arising from lifting work because based on their experience, those are some of the most frequent types of injury. In this case, an RA team can be specially formed and scoped to look into hand and finger injuries during crane lifting work. The RA will then form the basis for WSH management activities such as a campaign to raise awareness on the prevention of hand and finger injuries during lifting.

During preparation, the RM/RA team will have to gather relevant information such as layout, workflow, manufacturer instructions, safety data sheets (for chemicals), etc. Different team members can assess different aspects of the collated information and then discuss them in meetings.

4.4.2 Risk assessment

After the preparation stage, RA will have to be conducted. This stage includes hazard identification, risk evaluation and risk control. During hazard identification, the RM/RA team will have to brainstorm for possible hazards. There are many hazard identification methods. One of the most basic is to review relevant hazards or safety issues documented in WSH audit reports, inspection data, incident investigation reports, injury and illness records, occupational health surveillance reports, first aid logs, observation records, employee feedback or consultation records and safety data sheets. These documents will supplement systematic hazard identification or RA techniques like Job Safety Analysis (JSA) / Job Hazard Analysis (JHA), What-if Analysis, Failure Mode and Effect Analysis (FMEA), Hazard and Operability (HAZOP) Study, Fault Tree Analysis (FTA) and Event Tree Analysis (ETA) (see Table 4.3). The wide range of hazard

Table 4.3 Risk assessment techniques

Technique	Description	Remarks
Job Safety Analysis (JSA)/Job Hazard Analysis (JHA)	JSA and JHA are very similar techniques, and both are conducted according to the following steps: • selecting the job to be analysed • breaking the job down into a sequence of steps • identifying potential hazards • determining preventive measures to overcome these hazards Some JSAs require the risk level to be determined, while JHAs usually do not require risk levels to be determined. This is the key difference between the two techniques, but many variants exist.	JSA and JHA are relatively simple to use. They are usually structured based on a table similar to the table recommended in the RMCP. However, JHA does not require risk levels to be identified. They are suitable if the focus is on an activity, process or task that has specific steps. They are the most commonly used techniques in the construction industry.
What-if Analysis	A What-if Analysis consists of structured brainstorming to determine what can go wrong in a given scenario. The person performing the analysis then judges the likelihood that things will go wrong and considers the consequences. What-if Analysis can be applied at virtually any point in the evaluation process. Based on the	What-if Analysis is essentially a brainstorming session to identify hazards and assess their risk levels. People involved must be very experienced and the facilitator must be able to guide the group. As compared to the JSA/JHA approach, which is

	answers to what-if questions, informed judgments can be made concerning the acceptability of those risks. A course of action can be outlined for risks deemed unacceptable (adapted from https://www.acs.org/). The what-if questions should be generated prior to the brainstorming session and a group of experts must be involved in the analysis.	guided by a detailed template, the what-if analysis is comparatively unstructured.
Failure Mode and Effect Analysis (FMEA)	The FMEA is also a template-based approach, and it has a long history in reliability engineering. FMEA typically takes a piece of machinery or equipment and divides it into sub-systems that can be assessed effectively. The process involves: (1) Identification of the component and parent system, (2) Failure mode and cause of failure, (3) Effect of the failure on the sub-system or system and (4) Method of detection and diagnostic aids available. FMEA is quantitative in nature and it estimates the risk levels based on failure and error data collected by plants and manufacturers.	FMEA is commonly used in the manufacturing industry. Its quantitative nature makes it feasible only in situations where the data are readily available. Nevertheless, it is still possible to use it based on qualitative assessments.

(Continued)

Table 4.3   (*Continued*)

Technique	Description	Remarks
Hazard and Operability (HAZOP)	HAZOP studies have been used for many years as a formal means for the review of chemical process designs. A HAZOP study is a systematic search for hazards which are defined as deviations within parameters that may have dangerous consequences. In the process industry, these deviations concern process parameters such as flow, temperature, pressure etc. (adapted from www.hse.gov.uk).	During a HAZOP study, the team will typically spend a few days to a few weeks evaluating the piping and instrumentation diagrams of a plant to be constructed. The facilitator will use a set of guidewords (like more, less, high, low, fast, slow) to match with the process parameters so as to brainstorm what are the possible effects of these deviations. Commonly used in industries like oil and gas, chemical and energy-related.
Fault Tree Analysis (FTA) and Event Tree Analysis (ETA)	A fault tree is a diagram that displays the logical interrelationship between the basic causes of the hazard. An FTA can be simple or complex depending on the system in question. Complex analysis involves the use of Boolean algebra to represent various failure states (adapted from www.hse.gov.uk). An ETA is a forward, bottom up, logical modelling technique for both success	FTA and ETA are commonly used in the nuclear and process industry to analyse complex engineering systems. Human errors can also be included to determine the reliability of these complex systems. FTA and ETA are quantitative in nature but can be adapted to be used qualitatively.

	and failure that explores responses through a single initiating event and lays a path for assessing probabilities of the outcomes and overall system analysis. This analysis technique is used to analyse the effects of functioning or failed systems given that an event has occurred. (adapted from https://en.wikipedia.org/wiki/ Event_ tree_analysis)	
Bowtie Analysis	The bowtie method is an RA method that can be used to analyse and communicate how high-risk scenarios develop. A bowtie gives a visual summary of all plausible risk scenarios concerning a certain hazard that could occur. By identifying control measures, the bowtie displays what a company does to control those scenarios. Control measures are also known as barriers. They are elements we have in place to ensure the hazards we deal with are kept in a wanted state. Barriers can be systems, regulations, design aspects, and so on (adapted from Bowtie Methodology Manual by CGE Risk Management Solutions B.V.).	The bowtie is a qualitative method that relies on a set of symbols to summarise a series of hazards and controls into diagrammatic form. The bowtie requires specialised software to be implemented effectively as the number of symbols and diagrams will grow very quickly. However, each bowtie diagram is a good summary of the hazards and controls that a particular activity has.

identification and RA techniques can be confusing, but in Singapore, most organisations use the approach recommended in the Risk Management Code of Practice (RMCP), which is essentially an activity-based RA (similar to JSA/JHA) or a trade-based RA. The activity-based RA and trade-based RA will be elaborated later in this chapter. In addition, if the workplace already exists, an inspection or walkabout can be conducted as part of the RA. Furthermore, during these walkabouts, checklists can be used to help RA teams identify hazards and observations, and interviews can also be conducted to obtain valuable information about how existing controls are performing. In general, the following categories of hazards should be considered:

- physical (e.g. fire, noise, ergonomics, heat, radiation);
- mechanical (e.g. moving parts, rotating parts);
- electrical (e.g. voltage, current, static charge, magnetic fields);
- chemical (e.g. flammables, toxics, corrosives, reactive materials);
- biological (e.g. blood-borne pathogens, virus); and
- psychosocial (e.g. stress, fatigue).

Human and behavioural factors are important aspects that must be considered during RA. Human factors literature (e.g. Health and Safety Executive, 1999) identified two broad types of human failures: errors and violations (see Figure 4.6). The key difference between errors and violations is that violations are deliberate and errors are unintentional. However, both involve an individual not acting in accordance to accepted standard, rule or procedure.

Skill-based errors are further classified into slips of action and lapses of memory. Slips of action or slips happen when an individual performing a very familiar task performs the wrong action. Using driving as an example, slips can involve stepping

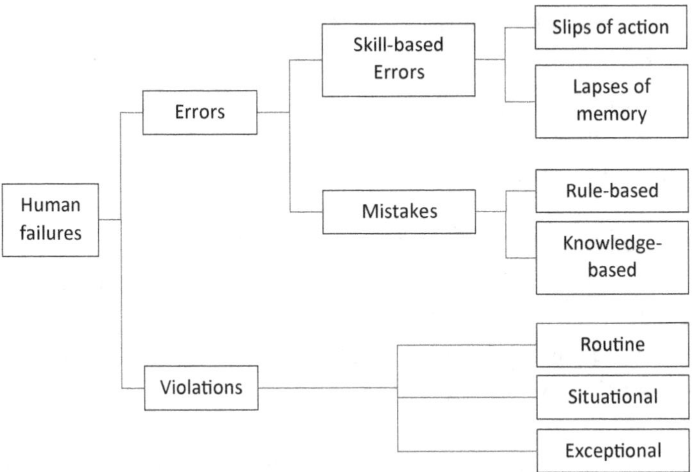

Figure 4.6   Types of human failures

on the brake too late, not checking on the blind spot, stepping onto the accelerator too hard, making a wrong turn, and stepping onto the accelerator when the intention is to brake. Lapses of memory or lapses happen when an individual performing a very familiar task forget to do an action, got lost during a series of actions, or forget the purpose of the actions.

Mistakes involve the failure of mental processes during planning, assessing, forming of intentions and judging consequences. Rule-based mistakes occur when we use familiar rules that are inappropriate for the situation. For example, when an engineer inspecting a faulty crane assumes that certain components that are indicated as "maintenance free" by other manufacturers also apply to the crane that he is inspecting. He thus fails to inspect certain components, resulting in a crane accident later on. Knowledge-based mistakes occur in unfamiliar situations when an individual has to diagnose the situation to select the right action, which is error-prone. An example of knowledge-based mistakes is when a construction work crew detects cracks on the supporting structures during pouring of concrete—and as the company does not have a pre-determined monitoring and evacuation procedure, the site personnel have to discuss the appropriate actions and misdiagnosed the cracks; and while doing so, fail to evacuate the workers, resulting in a major accident later on.

Some of the key factors influencing occurrence of human errors include presence of work environment stressors (e.g. heat, poor lighting, noise, and confined space), excessive task demands (e.g. high workload, unrealistic expectations on alertness, repetitive and monotonous tasks, excessive distractions), social and organisational stressors (e.g. insufficient human resource, demanding shift arrangement, conflicts between colleagues, peer-pressure and work bullies), individual stressors (e.g. lack of knowledge, fatigue, non-work problems, and misuse of alcohol or drugs), and equipment stressors (e.g. poorly designed user interface, and unclear instructions). These factors must be considered during RA.

Violations are another aspect of human failures that must be considered during RA. Many accidents occur due to violations and it is necessary for every RA to consider the tendency for workers to take shortcuts or violate SWPs, safety rules, and risk controls. Some possible violations include by-passing safety checks, disabling machine safety features, not using or misusing PPE, and unauthorised use or misuse of equipment. Violations can be routine, situational or exceptional. Routine violations are an indication that breaking safety rules is the norm in an organisation and is a reflection of the detrimental state of safety culture in that organisation. Situational violations happen when there are certain stressors arising from the individuals, work environment, equipment, job, social context or organisation, that appear to justify the violations. Exceptional violations happen only during emergencies or rare events, where emergency procedures or SWPs were violated based on the wrong belief that the benefits of the violation outweigh the cost of following them.

Likelihood / Severity	Rare (1)	Remote (2)	Occasional (3)	Frequent (4)	Almost certain (5)
Catastrophic (5)	5	10	15	20	25
Major (4)	4	8	12	16	20
Moderate (3)	3	6	9	12	15
Minor (2)	2	4	6	8	10
Negligible (1)	1	2	3	4	5

RPN 1–3 = Low Risk; RPN 4–12 = Medium Risk; RPN 15–25 = High Risk

Figure 4.7   Risk matrix (Workplace Safety and Health Council, 2015)

Table 4.4   Severity level

Level	Severity	Description
5	Catastrophic	Death, fatal diseases or multiple major injuries.
4	Major	Serious injuries or life-threatening occupational diseases (including amputations, major fractures, multiple injuries, occupational cancers, acute poisoning, disabilities and deafness).
3	Moderate	Injury or ill-health requiring medical treatment (includes lacerations, burns, sprains, minor fractures, dermatitis and work-related upper limb disorders).
2	Minor	Injury or ill-health requiring first-aid only (includes minor cuts and bruises, irritation, ill-health with temporary discomfort).
1	Negligible	Negligible injury.

When considering human factors during RA, the RA team must also consider individual health risk factors, including medical health issues, smoking, and alcohol misuse.

The risk matrix in Figure 4.7, together with the severity levels and likelihood levels in Tables 4.4 and 4.5, respectively, are described in the RMCP. They are used to evaluate hazards and assign a risk priority number (RPN). Once the RPN is determined,

Table 4.5   Likelihood level

Level	Likelihood	Description
1	Rare	Not expected to occur but still possible.
2	Remote	Not likely to occur under normal circumstances.
3	Occasional	Possible or known to occur.
4	Frequent	Common occurrence.
5	Almost certain	Continual or repeating experience.

the risk level can then be determined, and the RM team can then refer to Table 4.6 for the recommended actions.

Figure 4.7 and Tables 4.4, 4.5 and 4.6, which are taken from the RMCP, are guidelines, but they represent industry norms and generally accepted levels of risk tolerance. Thus, the recommended risk matrix in RMCP and other national codes of practice should be considered the default risk matrix for the relevant countries, unless the organisation has better alternatives. Nevertheless, organisations can create their own risk matrix, severity and likelihood levels, and recommended actions, and calibrate them to reflect their risk appetite and tolerability. However, any deviation from the guidelines provided in the RMCP should be conservative, i.e. the appetite for risk would usually be less than the national appetite. One way to calibrate risk martix and levels is to identify common hazards that most employees of the organisation are familiar with. Next, sample employees who are reflective of the population to assess those hazards individually. Subsequently, the RPNs generated by the sampled employees can then be used to assess if the definitions in Tables 4.4 and 4.5 are suitable. The higher the RPN, the more attention and control are required. The hazards assessed to be in the top right corner (where the RPN = 15 to 25) need to have additional risk controls before work can be conducted. Controls can be identified based on the hierarchy of control and Haddon's countermeasures. The diagonal area of the matrix (where RPN = 4 to 12) indicates that the risk level is

Table 4.6   Recommended actions for risk levels

Risk Level	Risk Acceptability	Recommended Actions
Low	Acceptable	• No additional risk control measures may be needed. • Frequent review and monitoring of hazards are required to ensure that the risk level assigned is accurate and does not increase over time.
Medium	Tolerable	• A careful evaluation of the hazards should be carried out to ensure that the risk level is reduced to ALARP within a defined time period. • Interim risk control measures, such as administrative controls or PPE, may be implemented while longer term measures are being established. • Management attention is required, which means management must have a monitoring system in place so that they can stop work and evacuate its workers if the workplace becomes unsafe.
High	Not acceptable	• High Risk level must be reduced to at least Medium Risk before work starts. • There should not be any interim risk control measures. Risk control measures should not be overly dependent on PPE. • If practicable, the hazard should be eliminated before work starts. • Management review is required before work starts.

tolerable (medium), but additional control measures should be implemented whenever reasonable and practicable. The lower left zone (where RPN = 1 to 3) indicates that the hazard is low-risk and generally acceptable, but practicable controls that can push the risk lower should still be implemented. The different zones are aligned with the HSE Framework for Tolerability of Risk (Figure 4.4).

An RA is meant to be a live document. As highlighted in the WSH (RM) Regulations require the employer, self-employed and principal are required to review the RA where there is a "significant change in work practices or procedures" (see Regulation 7(2)(b)). The RMCP, clause 4.2.1.12, also indicates that "upon any accident, incident, near miss or dangerous occurrence;" and "when new information on WSH risks is made known" the RA must be reviewed and, if necessary, revised. An RA review is similar to an RA, but instead of starting from scratch, the RA Team will review existing RAs based on recent incidents, changes and/or additional information.

Even though RA is a systematic process that should lead to "consistent and reliable results" (RMCP clause 6.2.5), it is only useful if the employer, self-employed or principal and its RA Team are committed to safety and conduct the RA carefully, meticulously and seriously. If the employer, self-employed or principal and its RA Team treat RA as a "paper exercise" only to satisfy requirements, the RA will not be effective and will not have significant benefits on WSH. Furthermore, the RA Team is not meant to work independently. They should keep the RM Team and management updated, and seek their approval and support whenever necessary.

Another important consideration during RA is highlighted in clause 4.3.1.4.4 of OHSAS18002:2008 Occupational health and safety management systems—Guidelines for the implementation of OHSAS 18001:2007 (British Standards Institute, 2008), which states that (emphasis inserted), "Hazards that could cause harm to large numbers of persons should be given careful consideration *even when it is less likely for such severe consequences to occur*". This indicates that the RA team should always be cautious and give more weight to consequences so as to effectively prevent rare major accidents. Major accidents, a type of "black swan" event, are unlikely events based on their frequency of occurrence, that have very severe consequences. The RA team must be mindful that they do not have full knowledge of hazards, risk (especially likelihood) and effectiveness of their controls. Therefore, it is critical for RA teams to develop monitoring procedures and continuously update the RA whenever there are changes to work practices, incidents, near hits, and when new WSH information are available. An example is the Nicoll Highway collapse, where information about waler beam buckling and abnormal instrumentation readings were not carefully evaluated as part of an RA review and the project team continued to assume that a tunnel collapse is not possible, possibly because tunnel collapses had never occurred in Singapore prior to the accident. It must be noted that black swan events typically appear to be rare or extremely unlikely to organisations because of lack of information or knowledge, or failure to systematically consider available information. Thus, by requiring a structured and on-going RA process, organisations will be less likely to miss critical WSH information and can guard against major accidents.

4.4.3 Implementation and review

To facilitate implementation, at the end of the RA, there should be an RM plan to ensure that all stakeholders are aware of the key hazards, their risk levels and the corresponding control. The RM plan should define the scope of the RM, the company job functions, the RA methodology, the risk control measures and the schedule or programme for implementing the RM plan. Most importantly, the RM plan should contain the actions that are required to implement the different controls identified during the RA. Many a time, stopgap measures (usually lower on the hierarchy of control, i.e. administrative controls and PPE) need to be implemented first because more effective measures need time for implementation, e.g. installing an engineering control. These details need to be spelled out in the RM plan such that the risk level during operations is at most tolerable or medium. The plan should clearly stipulate:

- what controls have to be implemented
- how to implement the controls
- who are to implement the controls
- where to implement the controls
- when to implement the controls

During implementation, it is important for organisations to ensure buy-in from supervisors and workers through communication and consultation. In terms of communication, it is critical to ensure that the nature of the hazards, risk levels and controls are explained. The relevant safe work procedures should be communicated in a suitable format to the relevant stakeholders, e.g. in pictures or comics to overcome possible communication barriers. The controls will impact the work environment, equipment, work procedures and personnel. Thus, the impact of the changes should be clearly communicated before implementation.

For example, the RA team for the installation of a dry wall along the edge of a mezzanine floor in a building was in a rush to finish the RA. They identified the hazard of falling over the edge of the mezzanine floor (3 m above the ground floor) in a new workshop and noted that due to the lack of height clearance, lanyard and personal energy absorbers will not be suitable in protecting the workers. Thus, they recommended the use of safety or anti-fall nets to be installed along the perimeter of the mezzanine floor. However, the nets made it infeasible for the scissors lifts to be used to assist in the installation of the dry wall. Workers familiar with the work should have been able to identify the problem easily but were not consulted because the RA team conducted the RA in a rush. This resulted in delays in the project as a new RA had to be conducted and the installed safety net had to be removed.

To facilitate communications, the organisation should clearly establish the different channels of communication, including WSH committee meetings, feedback sessions, small group meetings (e.g. tool-box meeting, and shift handover meeting), one-on-one discussion, email or online messaging, telephone calls and notice board/bulletins. The stakeholders that must be involved in the development and communication of the RM plan can include senior management, supervisors, subject matter experts, workers, consultants, contractors, and suppliers. The effectiveness of the controls and risk of hazards must be monitored during implementation. The monitoring process must be carefully considered during RA, where the RA team, with inputs from relevant experts, must consider possible sources of uncertainty or inaccuracy that could render the RA unreliable. These uncertainties can be due to lack of information (during RA) about the work, workers, environment, equipment, stakeholders, and risk controls. The monitoring system can be based on inspections, sensor or instruments, such as CCTV or other technology, but it should be suitable and proportionate to the work situation.

RA reviews should be conducted at least once every three years, whenever there are significant changes to the work, when there are incidents, near-hits or dangerous occurrences, or and when new information on WSH risks is made known. The RM plan should also clearly define the communication, consultation and review activities.

4.5 Common Errors in Risk Management

Despite its importance, in actual implementation, many RM and RA processes are saddled with errors. Table 4.7 summarises the common errors identified in the document, "E-Fact 32 — Common errors in the risk assessment process" (European Agency for Safety and Health at Work, 2008), adapted based on the author's experience.

Table 4.7　Common Errors in Risk Management

RM Stage	Common Error	Remarks
Preparation	• Not involving a team of people in the assessment	• RM and RA should be a team effort, where different opinions and ideas are collected to ensure effectiveness (e.g. comprehensiveness of hazards identified, representativeness of risk levels assigned, and proportionality and suitability of risk controls selected).
	• Not including people with the right expertise and experience	• Expertise can include technical or specialist knowledge of the work process, environment, material and equipment used. • Experience refers to the practical experience of carrying out the work; experienced frontline workers and/or supervisors should be involved.
	• Not designating the RA to a person who is competent	• RA process requires competencies such as facilitation skills, knowledge of RA techniques, and ability to communicate with different stakeholders.
	• Involving experts (including consultants) in the RA process who are not familiar with the organisation	• It is useful to involve experts with technical expertise, but the RM/RA team need to orientate them to the RM/RA process so that the experts understand the relevant WSH policies and procedures, the RM/RA method used, the scope of their engagement, the role of the RM/RA team and the resources available to the expert. • It is important that the experts work with the RM/RA team to conduct the RA; if the experts conduct the RA without any

		involvement from the RM/RA team, the RA process may become impractical and there may be limited buy-ins, therefore causing problems during implementation.
	• Not allocating enough time and resources to the RM/RA process	• RM/RA process is a planning process that should be given enough emphasis, resources and support from management; management frequently underestimates the time and resources needed for a thorough RM/RA process.
	• Not collecting enough background or site-specific information	• RM/RA team may miss out required information during preparation, e.g. site-specific characteristics (e.g. traffic condition and community profile), information on specific type of equipment to be used (including age of equipment), and documentations on past incidents and ill-health.
Risk Assessment: Hazard Identification	• Overlooking possible hazards and incidents	• RM/RA team may miss out hazards that they are not familiar with such as health risks (e.g. psychosocial issues and chronic diseases). • RM/RA team may assume that hazards and incidents that had never happened before will never happen in the future, therefore causing hazards that are of low likelihood, but high severity to be overlooked; it is important for hazard identification to be comprehensive and not screen out possible hazards too early; the filtering and prioritisation should only be done during risk evaluation.

(Continued)

Table 4.7   (*Continued*)

RM Stage	Common Error	Remarks
		• Inappropriate use of checklists can cause hazards to be overlooked; checklists are usually the first steps to a hazard identification and RM/RA teams will need to go beyond the checklists to ensure that the hazards identified is comprehensive. • Hazards can be overlooked if the work process is not carefully documented; non-routine tasks and activities like maintenance, repair and remedial or corrective actions when foreseeable errors were made need to be included as part of the work and task inventory. • Hazards arising due to interactions between different work activities due to their proximity, sequence or dependencies can be overlooked because the RM/RA team was focusing on each of the activities separately.
	• Overlooking other people exposed to the hazards	• Hazards may be identified, but other people exposed to the hazards, such as other employees not directly involved in the work, subcontractors, visitors and members of the public, may not be adequately identified.
	• Overlooking vulnerability of specific groups of people exposed to the hazards	• The RM/RA team typically focuses on the "average worker" exposed to a hazard, but specific groups of workers, e.g. older workers, pregnant or workers with disability, may be more susceptible to certain hazards and may require additional control measures.

	• Overlooking specific characteristics of the work being assessed that differs from the past	• The RM/RA team may be over-relying on past RAs or past experience; it is important to customise the RA by considering the specific characteristics of the current work, e.g. location, work methods, workers profile, equipment, material andstructure; it is important to consider the changes made since the last RA and not assume that everything is business as usual.
	• Assuming that a hazard will be looked into by others	• RM/RA team assumes that another RM/RA team, engineering team or other relevant teams is/are looking into a hazard and removes the hazard from the team's scope of work.
Risk Assessment: Risk Evaluation	• Not considering long term consequences	• Long term consequences, especially health-related consequences, may not be assessed.
	• Lack of consistent description of hazard and possible incident	• The RM/RA team may not develop a coherent description of the hazard and possible incident so that the risk levels considered by different RM/RA team members are based on the same set of descriptors. • A useful approach is to assess all hazards based on the concept of "worst credible scenario", where a reasonable estimation of the worst-case scenario is used to ensure consistency across the risk evaluation.

(Continued)

Table 4.7   (*Continued*)

RM Stage	Common Error	Remarks
	• Not considering reliability and effectiveness of risk control measures when assigning risk level	• When considering the severity and likelihood of a hazard and its potential consequences, the RM/RA team might neglect to consider the likelihood of the risk control not performing as intended due to issues such as equipment failure, human errors, and violations.
	• Reducing severity, when likelihood should be reduced	• When additional controls are implemented, the severity should be reduced only if the controls are reliable in reducing the energy or amount of hazardous substances that a worker will contact during an incident. • Using the "worst credible scenario" rule, severity is less likely to be reduced when compared with likelihood.
	• Not doing validity check on the risk levels assigned	• RM/RA team should sort all hazards based on the risk level and check if the risk levels assigned make sense when compared across the hazards. • Another approach is to compare with benchmark risk levels that most people can relate to, e.g. risk level of crossing the road, driving and taking public transport.
Risk Assessment: Risk Control	• Not taking the hierarchy of control into consideration	• RM/RA team might think that elimination is not possible and did not even brainstorm on possible opportunities to eliminate hazards.

		• RM/RA team might over-rely on administrative controls (e.g. training, warning signs and briefings) and personal protective equipment, and do not strive to identify substitution and engineering controls.
	• Assume that transferring the risk means no risk	• Risk transfer is also known as risk sharing; when contractors or consultants are brought in for a specific scope of work, the principal engaging the contractor or consultant will have to work with them to manage WSH risks and not assume that they have transferred all their risk and no longer have any WSH duties.
	• Not prioritising the controls, clearly assigning employees overseeing the implementation of the controls and stipulating due dates	• Due to the wide range of hazards identified, it is not possible to implement all the additional controls immediately; thus, it is important to prioritise the controls based on the inherent risk levels (risk level before additional controls), and clearly identify the employees to ensure implementation of the risk controls with strict due dates. • Special attention should be given to controls for hazards with high severity, but low likelihood; these hazards can cause catastrophic consequences and their prevention are dependent on reliable controls that are closely monitored. • The due dates for implementation should be tied to start date of relevant task or activity.

(Continued)

Table 4.7 (*Continued*)

RM Stage	Common Error	Remarks
Implementation and Review, and Communication	• Not involving or communicating with workers	• Implementation of risk controls is heavily dependent on buy-in and cooperation from workers; it is important that there are communication and consultation with workers on the risk controls and give them opportunities to feedback on the controls. • Workers must feel safe to work and communication affects their perception of risk; therefore, management must continue to assure workers of the effectiveness of the risk controls and encourage hazard reporting if any of the controls is in doubt.
	• Not reviewing the RA when necessary	• A RA review must be conducted at least once every three years, but when there are significant changes to the work, when incidents arise and when new WSH information becomes available, the RA should be reviewed. • Organisations and individuals may assume that certain work changes, incidents and WSH information are not significant enough to trigger a RA review; on the contrary, RA reviews should be expected because RAs are frequently conducted prior to the beginning of work, so it is frequently inaccurate; thus as more information becomes available, RA reviews must be conducted.

		• Management should make RA reviews a norm and a natural part of work, especially highly dynamic and project-based work, e.g. construction, and shipbuilding and ship repair work.
	• Not monitoring hazards, risk levels and risk controls	• During implementation, the organisation must continue to monitor the hazards, risk levels and risk controls to assure that the risk levels do not become unacceptable unknowingly.
	• Not implementing the risk controls in the RA	• Organisations may treat the RA as a "mere paper exercise" for the sake of satisfying regulatory requirements (i.e. surface compliance), and the risk controls are not truly implemented in the workplace. • Organisations with poor safety culture would typically have such behaviours; it is up to management to proactively check that RA is given emphasis and attention so that frontline employees are motivated to implement the risk controls.
Keeping Record	• Not recording RA	• RA is meant to be a deliberate and structured approach, which prevents organisations and individuals from making intuitive judgement that are error prone and risky; the process of having the RA recorded is meant to facilitate the deliberate and structured approach.

(Continued)

Table 4.7   (*Continued*)

RM Stage	Common Error	Remarks
		• Information regarding hazards, risk levels and risk controls should also be carefully recorded to ensure that critical WSH-related information are considered. • Recording also helps to ensure that information is clearly communicated to relevant personnel, provides basis for RA reviews, faciliates tracking of status of implementation, and facilitate demonstration of due diligence. • Larger organisations with multiple RAs should consider using RM software to help them with RM and RA records; the RM software should help to structure the different RAs to minimise repetitive information from being documented in different records; the software should facilitate cross-referencing and updating of documents systematically.

Besides the common errors described in Table 4.7, it must be noted that risk perception is affected by many cognitive biases such as (based on Dobelli, 2013):

- Base rate neglect: A disregard of fundamental frequencies of different types of accidents and illnesses. During RA, organisations fail to consider the overall accident frequencies in the industry and assign unrealistic likelihoods to their own workplaces.
- Over-confidence: When individuals are asked to rate their competency or likelihood to get into an accident, most will over-rate their competency in relation to others and assume that they will not get into an accident.
- Availability bias: Most people do not get into an accident or illness on a daily basis. Thus, based on their experience, the most available memory is those of non-accidents or days with no occupational illness. This can lead to developing a mindset biased to the assumption that accidents and illness will not happen in their workplace.
- Survivorship bias: Similar to the above biases, we tend to focus on success or 'non-failure'. As Dobelli puts it, "People systematically overestimate their chances of success. Guard against it by frequently visiting the graves of once-promising projects, investments and careers. It is a sad walk, but one that should clear your mind."

To deal with these biases, it is important to communicate WSH information, statistics and case studies to the managers, supervisors and workers. Awareness of these cognitive biases will also help to reduce their potential impact.

Another fundamental problem common in organisations is the failure to implement the RA due to a disproportionate focus on productivity and poor communication. These issues are related to the safety culture of the organisation. Some organisations are over-focused on the documentations needed in RM.

They think that having the set of documents means that they are protected from possible prosecution by the authorities, rendering RM or RA a mere paper exercise. However, paperwork without implementation is still poor management and the paperwork takes up time without adding value. If accidents happen, the organisation will still be heavily penalised.

Communication should always be focused on the receiver's preferences and suitability for the mode of communication. For example, the use of pictures and videos is much more effective for workers and supervisors than lengthy RA documents. In industries where workers of different nationalities are present, the clarity of communication needs to be looked into very carefully. The person conducting the briefing must prepare beforehand, conduct the briefing clearly and check if the receivers of the information have understood the information.

4.6 Systems Thinking and Risk Management

Lee and Green (2015) highlighted how systems thinking concepts can be used to improve enterprise RM. Since the processes and goals of RM is similar across different fields, the insights from Lee and Green (2015) are applicable to WSH RM. The following highlights some useful insights adapted from Lee and Green (2015).

First, systems thinking emphasises the whole and not single components. During RM processes, a divide and conquer approach is taken to structure and scope different RAs. This may inadvertently cause hazards that arise from the interactions between different system components and processes to be missed. Therefore, management must ensure that WSH managers or risk managers are assigned to maintain a holistic view of the RM processes to check that interfaces, interdependencies and "grey areas" are being studied by specific RM/RA teams.

Second, systems thinking recognises that mental models and organisational culture are critical components of any system. To ensure effective RM, the management must recognise that risk is a human construct and it is important to understand the mental models and organisational culture that influence risk perceptions. This is where risk communication must consider the mental models of stakeholders and organisational culture, and plan and target the risk communication activities accordingly.

Third, it must be acknowledged that the overall system objectives (e.g. zero accident), may conflict with individuals' goals or other system objectives. This is where management must consciously identify such conflicts or the resulting resistance so as to assure the effectiveness of RM processes.

Lastly, double loop learning (which is similar to the concept of underlying factors in Chapter 3) is more effective in the continuous improvement of WSH RM than single loop learning. In the context of WSH RM, single loop learning is about adjusting WSH objectives and approaches by comparing the current performance with established objectives. This can include the objective of implementing a series of risk controls and comparing that with the actual implementation on the ground. If the objective of implementation is not achieved, management can continue to increase enforcement or communication. Such single loop learning is the typical control loop used in most organisations. Double loop learning is focused on evaluating the underlying assumptions related to WSH RM when comparing WSH performance against objectives. Using the same example, when risk controls are not implemented as per plan, management should seek to understand what are the underlying mindset, cultural, systemic or leadership issues that could be hindering the implementation. These systems thinking management principles are useful guidelines for managers overseeing RM.

4.7 Summary

RM is not unique to WSH management and it can be found in different management areas. RM is the cornerstone of WSH management and any workplace that is seeking to prevent accidents and ill-health must understand RM and implement it effectively. The RM process is a structured process of preparation, RA, implementation, review recording and communication. It helps organisations to proactively reduce risk. However, there are several common errors in implementing RM and the management must be aware of them. The chapter also described how systems thinking's focus on the holistic, and a fundamental understanding of system behaviour, provide useful management guidelines for all WSH RM efforts.

Review Questions

1. Why is RA the cornerstone of the WSH Act?
2. What are the factors that affect risk perception?
3. Describe the roles of the following people in RM: Employer, Manager, Human Resource Manager, RM and RA Leaders, and Employees.
4. When can an organisation stop monitoring a specific hazard?
5. Describe six categories of hazards.
6. Describe three examples of human and cultural factors that should be considered in an RA.
7. Using the recommended 5 × 5 matrix, what are some of the high-risk, medium-risk and low-risk hazards in your industry? How do you justify your answer?
8. What do you think are some of the possible problems in implementing the RMCP in a small workplace?
9. What are some of the common errors in WSH RM?
10. What are some of the systems thinking concepts that can help to improve RM?

References

British Standards Institute (2008). "OHSAS18002:2008 Occupational health and safety management systems — Guidelines for the implementation of OHSAS 18001:2007." London.

British Standards Institute (2009). "PD ISO Guide 73:2009 Risk Management — Vocabulary." London.

British Standards Institute (2018). "BS ISO 31000:2018 Risk management — Guidelines." London.

Committee on Safety and Health at Work (1972). "Safety And Health At Work: Report Of The Committee, 1970–72 (Roben's Report)". London: H.M. Stationery Office. Available at: http://www.mineac-cidents.com.au/uploads/robens-report-original.pdf (Accessed June 24, 2020).

Dobelli, R. (2013). *The Art of Thinking Clearly: Better Thinking, Better Decisions*. London: Hodder and Stoughton.

European Agency for Safety and Health at Work (2008). "E-Fact 32 — Common errors in the risk assessment process." Available at: https://osha.europa.eu/en/publications/e-facts/e-fact32/view (Accessed on February 24, 2020).

Health and Safety Executive (1999). *HSG 48 Reducing Error and Influencing Behaviour*. 2nd Edition. London: The Stationery Office.

Health and Safety Executive (2001). *Reducing Risks, Protecting People: HSE Decision-making Process*. London: The Stationery Office.

Lee, L. S. and Green, E. (2015). "Systems Thinking and its Implications in Enterprise Risk Management", Journal of Information Systems, 29(2), 195–210.

Workplace Safety and Health Council (2008). *Technical Advisory for Working at Height*. Singapore: WSHC.

Workplace Safety and Health Council (2015). "Code of Practice on Workplace Safety and Health (WSH) Risk Management." Singapore: WSHC.

CHAPTER 5

Design for Safety

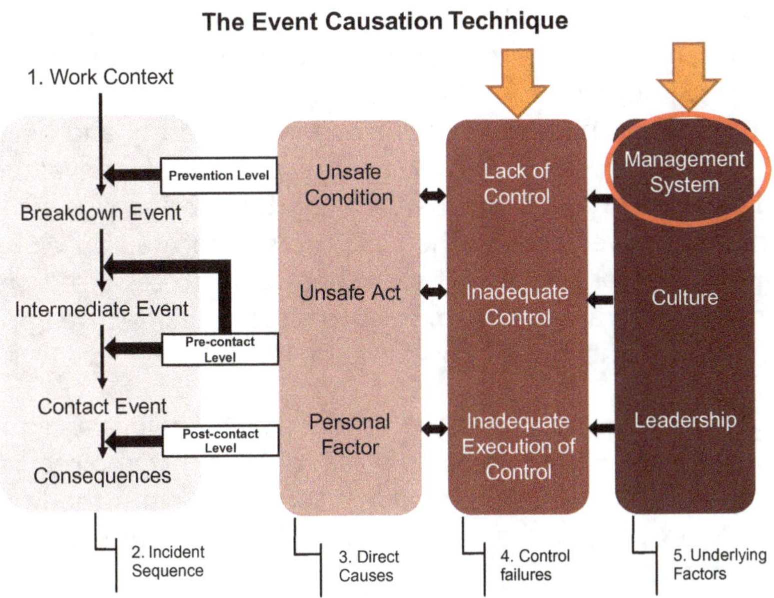

Design for Safety (DfS) is essentially a design risk assessment. It is an important aspect of the management system (see Event Causation Technique (ECT) diagram above). The end products of the DfS process are the design-related controls to prevent incidents downstream.

5.1 Introduction

Design for Safety (DfS) promotes early consideration of safety and health hazards during the design phase of a project, workplace or product. With early intervention, hazards can be more effectively eliminated or controlled, leading to safer products, workplaces, and production and construction processes. This chapter will focus on how DfS is practised in the built environment, but the processes and concepts will also be applicable to other industries.

DfS is practised in the built environment of many countries, including Australia, the UK, and Singapore. In Singapore, the Ministry of Manpower (MOM) enacted the Workplace Safety and Health (Design for Safety) Regulations 2015 (DfS Regulations) in July 2015, which were implemented from August 2016 onwards. Prior to the DfS Regulations, contractors have always been the main party involved in workplace safety and health (WSH). DfS imposes safety and health duties on the developers and designers, requiring them to conduct design risk assessments (RAs). Even though the DfS process is focused on WSH, it is also useful for the project team because it forces the project team to assess risks upstream so as to assure the effective implementation of the project downstream.

This chapter introduces the concept of DfS, relevant regulations and the Approved Code of Practice (ACoP). It also provides examples of how DfS can improve WSH throughout the lifecycle of a structure and discusses the challenges and success factors for DfS.

5.2 Defining Design for Safety

According to numerous studies, many fatalities in the construction industry can be attributed to design decisions or lack of planning (Behm, 2005; Workplace Safety and Health Council, 2015). Thus, the concept of DfS was introduced to minimise the risk of accidents and ill health through the consideration of hazards during upstream design phases of a construction project (Gambatese *et al.*, 2008). DfS is also known as prevention through design (López-Arquillos *et al.*, 2015), safe design (Safe Work Australia, 2018) and Construction (Design and Management) (Health and Safety Executive, 2015).

DfS is defined as: *"The practice of anticipating and "designing out" potential occupational safety and health hazards and risks associated with new processes, structures, equipment, or tools, and organizing work, such that it takes into consideration the construction, maintenance, decommissioning, and disposal/recycling of waste material, and recognizing the business and social benefits of doing so."* (Schulte *et al.*, 2008) (p. 115). Essentially, DfS is the RA of a design to identify the hazards that can pose safety and health risks at different stages of development of a workplace, building or facility, so that the hazard can be eliminated or mitigated through options such as design changes, risk control measures and the sharing of information.

The DfS process requires the involvement of all stakeholders. In the construction industry, the stakeholders include the client or developer, the designers, and the contractors. However, one important stakeholder not mentioned in the DfS Regulations is facilities management, who holds important WSH knowledge of maintenance and operation. Similarly, in other industries, the owner or client will usually engage designers and contractors to design and manufacture or construct an asset, equipment, machinery or product. This asset, equipment, machinery or product can be a vessel, a metalworking plant, a school building, a photocopier or a desk. Some industries, for example the oil and gas industry, have a very rigorous process to ensure safe operation and construction. This process uses RA methods like Process Hazard Analysis (PHA), Hazard and Operability (HAZOP), What-if Analysis, and Layers of Protection Analysis (LOPA). The construction industry also has a thorough design review process. For example, in Singapore, the Building Control Act and its subsidiary regulations provide stringent requirements to ensure that competent designers are engaged to oversee and check designs. However, the Building Control Act is focused on major hazards such as the collapse of a structure and building fires, and does not adequately cover WSH issues like slips, trips and falls, noise-induced deafness and getting struck by vehicles. DfS Regulations cover all WSH incidents, including non-major incidents that involve the unsafe actions of individual workers, and has a distinctly wider focus than the Building Control Act.

The concept of DfS is not new to the built environment. For example, the Europeans have been implementing the concept since the 1990s. In the UK, the Construction (Design and Management) (CDM) Regulations has been revised several times and had seen some success. The South African Construction Regulations (2003) and the Australian Occupational Health and Safety Acts have also implemented the concept of DfS. In Singapore, the DfS Regulations was enacted in 2015 and enforced in 2016. DfS is aligned with one of the key principles of the WSH Act: "reducing risks at the source by requiring all stakeholders to eliminate or minimise the risks they create." Despite the differences between the operationalisation of the concept of DfS, all countries implementing DfS seek to minimise risk of accidents and ill health through the consideration of hazards during the upstream design phases of a construction project.

As seen in Figure 5.1, as a project progresses from inception to completion, the scope for change decreases and the cost of change increases. Since DfS involves design changes to eliminate hazards and the implementation of controls to mitigate WSH risks, as a project progresses, it becomes more difficult to implement these changes and controls to improve WSH. Therefore, it is important for DfS to be implemented early in the design phase. However, it is also not possible for DfS reviews to be conducted when there is insufficient information about the design. Thus, DfS reviews must be conducted near the end of each design phase, when it is still possible to implement changes. In addition, the DfS review process considers the lifecycle of the structure or product, i.e. it considers the safety and health of workers, occupants, people in the community, maintenance

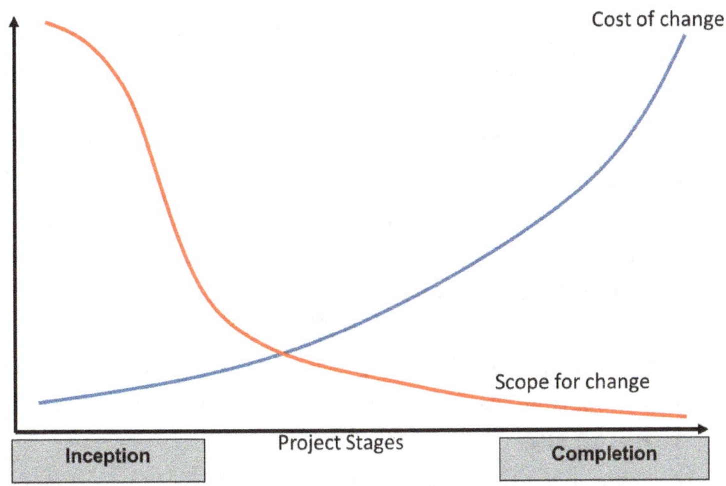

Figure 5.1   Cost of and scope for design changes to improve safety

workers, and demolition workers. Thus, it is useful to involve construction contractors and facilities managers in the design review process to ensure comprehensive hazard identification and suitable risk evaluation. The DfS process will be discussed in more detail subsequently.

5.3 Design for Safety Regulations

DfS Regulations are applicable to projects undertaken by a developer in the course of the developer's business, where the contract sum is S\$10 million or more and involves development under section 3(1) of the Planning Act (Cap. 232). According to the Planning Act, development is defined as, "the carrying out of any building, engineering, mining, earthworks or other operations in, on, over or under land, or the making of any material change in the use of any building or land." The Planning Act section 3(2) describes specific exclusions that are not classified as development and hence not covered by the DfS regulations. For example, the maintenance, improvement or other alterations of a building which do not materially affect the external appearance or floor area of the building, minor or preliminary works and such temporary use of land as may be declared by the competent authority, and the carrying out by any statutory authority of any works for the purpose of laying, inspecting, repairing or renewing any sewers, mains, pipes, cables or other apparatus. On the other hand, the Planning Act section 3(3) highlighted several specific inclusions, e.g.

- "(b) the use as a dwelling-house of any building not originally constructed for human habitation involves a material change in the use of the building"
- "(c) the use for other purposes of a building or part of a building originally constructed as a dwelling-house involves a material change in the use of the building"
- "(d) the demolition or reconstruction of or addition to a building constitutes development"
- "(e) the use for the display of advertisements of any external part of a building which is not normally used for that purpose involves a material change in the use of the building"

However, it must be noted that the regulations only apply to projects that had a designer appointed on or after 1 August 2016. The DfS Regulations consist of four parts: (1) Preliminary, (2) Duties of Developer, (3) Duties of Designer and Contractor, and (4) Miscellaneous. The following section discusses the duties of the different stakeholders.

5.3.1 Developers

The developers are expected to ensure that "foreseeable risks" are eliminated. If it is not "reasonably practicable" to eliminate the risks, then the "design risk" must be "reduced to as low as reasonably practicable" (ALARP) by reducing the risk at its source and using collective protective measures instead of individual protective measures. In relation to a structure, "design risk" means anything present or absent in the design of the structure that increases the likelihood that an affected person (e.g. construction workers, and occupants who work in the structure after construction, including maintenance workers and demolition workers) suffering bodily injury when constructing, working at, or demolishing the structure. The structure refers to any permanent or temporary structures and it includes the products and mechanical or electrical systems that are part of it. Some examples of structures can include a school building, a temporary earth retaining structure, a green façade on which a landscape worker works, and a warehouse that is being demolished. The concept of reasonably practicable has been discussed in Chapter 1.

An example of working at heights is used to illustrate what it means to reduce risk at source and what collective and individual protective measures are. A DfS review for a condominium building using volumetric construction methods identified that during construction, many construction workers need to work at heights to install the volumetric modules. Even though the risk is lower than traditional cast-in-situ methods, the team of installers and lifting personnel will still be exposed to the risk of falling from height. The current approach is to require the site workers to install their own fall arrest lifelines when the modules are being installed on site. This means that contractors will have to manage the risk downstream (not at source). The fall arrest system is an individual protective measure. Even though it is ideal to totally eliminate the risk of falling from height, it is not feasible with the current state of technology. Thus, following the principle of reducing risk at source and the use of collective protective measures, the design team decided that they will design a set of foldable guard rails that can be folded when the modules are being transported. The foldable guard rails can then be put into place when the modules are still at the ground level of the site. The worker can use platform ladders or scissors lifts to install the guard rails before they are being lifted. The risk is now greatly reduced because the risk control measure (foldable guard rails) are designed for upstream installation and put in place at the prefabrication yard and not at the construction site. The guard rails are a collective protective measure because, in contrast to a fall arrest system, it can protect many workers at the same time and do not require them to actively ensure that their own control measures are in place during work.

Developers are to ensure that designers and contractors are competent to perform their duties under the DfS Regulations. The traditional approach of ensuring competency is through training. Thus, developers will usually require designers and contractors to have personnel who have taken DfS-related and WSH-related courses. Another way to demonstrate competency is through past experience in implementing DfS review processes.

In addition, developers have to ensure that the project is properly planned and managed. The DfS Regulations specifically highlight the need for "sufficient time and resources" and "relevant information" to be provided to designers and contractors so that they can perform their statutory duties under the WSH Act. However, it is not clearly established what constitutes "sufficient time and resources". One possible approach to determine sufficient time and resources is to make reference to the time and resources allocated to past similar projects that were safely completed. More research is needed to provide benchmarks to evaluate the sufficiency of time and resources allocated.

Developers need to convene DfS review meetings to identify all foreseeable design risks and discuss how each of the risks can be eliminated or reduced. The DfS processes involved will be discussed in more detail subsequently. Developers will have to ensure that all relevant designers and contractors attend the DfS review meetings. A DfS register containing information and records of the DfS review meeting and residual design risk must be kept up-to-date and made available to designers, contractors and the regulator (i.e. WSH Inspectors from the Ministry of Manpower). The DfS register is a legal document that must accompany the structure even when the owner changes. The developer must inform any new owner of the nature and purpose of the DfS register.

The developer is allowed to delegate the duties related to the DfS review meeting and DfS Register to a "DfS Professional" (DfSP) — a person assessed by the developer to be competent to perform the duties. The WSH Council approved several industry associations, such as the Institution of Engineers Singapore (IES), the Singapore Institute of Architects (SIA) and the Singapore Contractors Association Ltd (SCAL), to conduct a two-day DfSP course to certify DfSPs. Participants are required to have significant experience in the construction industry. The course also requires participants to submit a report containing a DfS register for assessment.

The developer must provide all the relevant information necessary for the DfSP to perform the delegated duties. On the other hand, the DfSP must, as soon as reasonably practicable after each DfS review meeting, provide the developer with all relevant information on each foreseeable design risk identified at the meeting and how each design risk can be eliminated or reduced. In addition, the DfSP must, as soon as reasonably practicable after any information or record is added to the DfS register, provide the developer with an updated copy of the DfS register. It must be noted that the developer cannot delegate its general duties with regard to the implementation of DfS; which are ensuring competent designers and contractors are engaged, ensuring that the project is properly planned and managed with sufficient time and resources for designers and contractors, and ensuring designers and contractors have all relevant information.

From the duties allocated in the DfS Regulations, it is obvious that developers have an important role to play in DfS. This is a suitable approach because past research (Goh and Chua, 2016; Toh *et al.*, 2016) has shown that developers are an important source of motivation for designers to implement DfS.

5.3.2 Designers

In the DfS Regulations, a "designer" is defined as the person who prepares a design plan relating to a structure. "Design plans" include drawings, building information modelling, design details, specifications, materials and bills of quantities (including specifications of articles or substances) relating to a structure, and calculations prepared for the purpose of a design. This definition is very broad and can even include quantity surveyors and contracts managers who develop bills of quantities that influence the safety of the design.

While preparing the design plan, designers (in particular, engineers, architects, and contractors) must "as far as reasonably practicable eliminate all foreseeable design risks." If it is not possible to eliminate the design risk, then the designer must propose to the person who appointed the designer (e.g. developer and designer and contractor) a modification to the design plan that reduces the design risk to ALARP. As discussed earlier, the modification must take into account the principle of reducing risk at source and adopt collective protective measures instead of individual protective measures. Designers must provide all relevant information on the design, construction or maintenance of the structure to the person who appointed the designer.

Many designers do not have sufficient experience and knowledge of the downstream processes, e.g. construction, maintenance and demolition. This knowledge gap can be minimised by involving construction contractors and facilities managers early in the process. Concurrently, it is useful to train designers on WSH hazards and controls, so that they are better able to assess WSH issues during the design phase.

5.3.3 Contractors

Contractors must inform the person who appointed them of any foreseeable design risk that they know of. They must also ensure the designers and subcontractors engaged by them are competent and provide them with relevant information.

Contractors are heavily regulated in terms of site WSH management and have the most direct control over site safety. In contrast, DfS is more focused on developers and designers. However, in the case of design and build projects, the contractor performs both the role of a designer and a contractor. As discussed earlier, it is advantageous for contractors to be involved in the design as early as possible so that they can provide inputs on the design before it is frozen. Contractors also engage designers to design temporary structures. In this context, they will also have to ensure that the designers are versatile with the DfS process.

5.4 Design for Safety Process

This section discusses two approaches used to facilitate the DfS process, CHAIR and GUIDE, with attention on the GUIDE process detailed in the WSH guidelines — DfS (Workplace Safety and Health Council, 2016) (hereafter referred to as "DfS Guidelines").

5.4.1 CHAIR: Construction hazard assessment and implication review

The Construction Hazard Assessment and Implication Review (CHAIR) was created by WorkCover New South Wales (2001). CHAIR is essentially a customisation of the HAZOP method (see Chapter 4 for more details) for the construction industry. CHAIR aims to identify and eliminate hazards or minimise risk levels of the hazards in a design as early as possible. The process involves all key stakeholders, who gather together to reflect on the design from a health and safety perspective. CHAIR is implemented at two key junctures in the project lifecycle: during the conceptual stage of a design (CHAIR-1) and just prior to construction (CHAIR-2 and -3). CHAIR-1 is meant to review the concept design. CHAIR-2 is meant to review construction and demolition issues. CHAIR-3 focuses on maintenance and repair issues.

CHAIR involves the following steps:

1. Assemble a CHAIR study team (include all stakeholders).
2. Define the objectives and the scope of the study.
3. Agree on a set of guidewords/prompts to assist in the brainstorming process.
4. Partition the design (CHAIR-1, CHAIR-3) or construction process (CHAIR-2) into logical blocks of appropriate size.
5. For each logical block, use various guidewords to assist with the identification of safety aspects/issues.
6. Discuss associated risks and determine if the safety risk can be eliminated.
7. If the safety risk cannot be eliminated, determine how it might be reduced.
8. Assess whether the proposed risk controls (e.g. expected safeguards, etc.) are appropriate (i.e. is the risk ALARP).
9. Document comments, actions and recommendations — determine appropriate solutions for design issues still to be resolved.

Since HAZOP is known for its rigour in assessing the safety and operability of a design, typically for process plants, CHAIR is also a rigorous process. However, HAZOP has not been known to be well-used in the construction industry due to the difficulties in creating guidewords and the time taken to conduct a HAZOP study. A brief enquiry with a major Australian construction contractor and an experienced Australian occupational safety and health expert in the construction industry did not indicate that CHAIR is well-used in Australia. In addition, Bluff (2003) discussed that the Memorandum of Understanding (MOU) signed by 17 contractors in New South Wales (NSW), Australia, which provided the support for the development of CHAIR, did not have a significant impact on DfS in NSW. Bluff (2003) indicated that the challenges in implementing DfS are: too much focus on paperwork, failure to address safety in design by clients, design profession and principal, poor programming practices and unrealistic timeframes. It can be hypothesised that the usefulness of CHAIR as a DfS tool is highly dependent on the safety attitude of clients, designers and contractors, adequate planning and the provision of a reasonable timeframe.

5.4.2 Design for safety guidelines by Workplace Safety and Health Council, Singapore

The DfS guidelines (Workplace Safety and Health Council, 2016) describe a recommended process for DfS review and documentation. The overall DfS review process is based on the GUIDE acronym which represents:

1. **G**roup together a review team consisting of main stakeholders
2. **U**nderstand the design concept by looking at the drawings and calculations or have designers elaborate on the design
3. **I**dentify the hazards and risks that arise as a result of the design or construction method. The risks and hazards should be recorded and analysed to see if they can be eliminated by changing the design.
4. **D**esign around the hazards and risks identified to eliminate or mitigate the risk.

5. **E**nter all the information — including information on vital design changes, that would affect safety and health or remaining hazards and risks to be mitigated — into DfS register.

Hazard identification (step 3) and design change (step 4) should be iterated until it is not reasonably practicable to eliminate or mitigate the hazards. At the end of the GUIDE process, the residual risk level of hazards should be recorded and signed off by the project team. The developer or the appointed DfSP will have to ensure that the DfS review process is embedded into the project.

According to the DfS Guidelines (see Figure 5.2), the GUIDE process should be conducted at three different junctures, each with a different focus. The three design reviews are named GUIDE-1, GUIDE-2 and GUIDE-3. For a traditional design-then-build contract, a GUIDE-1 design review is conducted after the concept design is nearly completed, but the design still has room for amendments. It is meant to focus on the project's general location, traffic, type of buildings in the surroundings, and other general characteristics. If necessary, at each of the stages, more than one meeting should be conducted to review all the possible hazards arising from the design. GUIDE-2 is meant to focus on the detailed design, maintenance and repair.

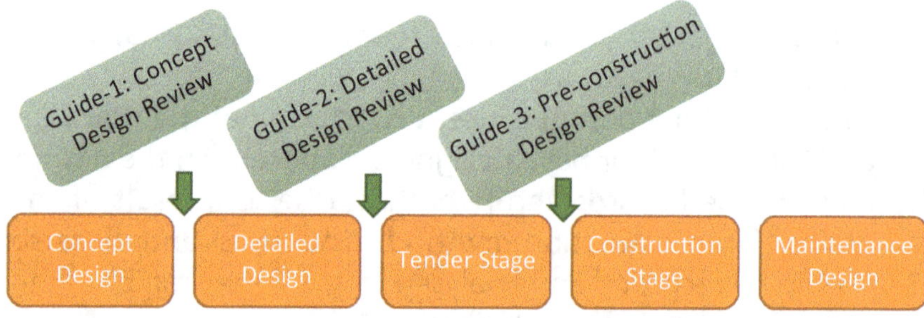

Figure 5.2   GUIDE process for design-then-build contracts (Workplace Safety and Health Council, 2016)

According to the DfS Guidelines, GUIDE-2 should review the hazards related to construction methods, access and egress and whether the design will create confined spaces or other hazards. Hazards during the maintenance and repair of the structure should also be reviewed.

Even though GUIDE-2 should include information provided by the contractor, this will be challenging for a traditional design-then-build project as the contractor may not be appointed yet. However, in alignment with approaches such as early contractor involvement (ECI), contractors can be engaged as consultants to improve the DfS review process. Since hazards during the maintenance and repair phase will be identified and eliminated or mitigated, it is also useful to involve experienced facilities managers and maintenance contractors during GUIDE-2.

A design-and-build (D&B) contract (see Figure 5.3) has the advantage of having the contractor and lead designers integrated into one. In the case of a D&B contract, the contractor will be able to provide construction method information early and influence the design to eliminate or mitigate risk at the construction stage.

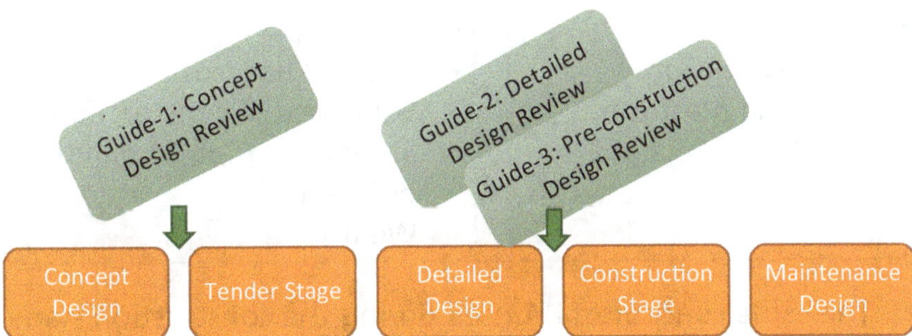

Figure 5.3   GUIDE process for design-and-build contracts (Workplace Safety and Health Council, 2016)

GUIDE-3 is focused on the pre-construction review, particularly on temporary works design and design by specialist contractors not covered during the concept and detailed design phases. Some of the key hazards and controls include: shoring, trenches and deep excavation, confined spaces and formwork and falsework. Residual risk of hazards and controls will have to be documented and highlighted to the contractors at the end of GUIDE-3. For example, if foldable guard rails are designed for the top of prefabricated modules to protect workers during installation, the folded guard rails need to be put into place by a site worker prior to the lifting of the modules. In this case, the residual hazard is falling from height when the worker is setting up the guard rails. The contractor will have to be notified of this hazard and will have to assess the risk during construction as well as put in place control measures to ensure that the risk is ALARP.

The DfS Guidelines highlight the following aspects that must be covered (but not limited to) during the GUIDE process:

• General design concept • Accessibility • Confined space • Emergency • Lighting • Excavation • Fall prevention • Working platforms • Hoisting or weight	• Layout • Maintenance • Material handling or storage • Means or methods • Operation • Physical hazards • Sequence of construction • Standardisation of building elements • Weather

The same concept of RA used during the construction stage can be used during the GUIDE design reviews. However, the design review is usually organised based on locations or design elements instead of activities. The hazards identified will then be controlled through elimination or mitigation. If mitigation is selected, then the hierarchy of control will also be used to guide the design risk control. Since the designers can make changes during design, they are expected to adopt more substitution and engineering controls and rely less on administrative controls and personal protective equipment (PPE).

The DfS Guidelines also provide a Safety and Health Risk Assessment Form which can be used to guide the DfS reviews. A similar form is provided in the Appendix to this chapter. As with any RA, the process must be structured and easy to administer. To ensure efficiency and effectiveness during the DfS review sessions, it is important to predetermine a list of design considerations that should be discussed during the design review meeting. These design considerations can be identified through a preliminary evaluation conducted prior to the design review meeting through the use of checklists, the use of guidewords (see CHAIR method discussed earlier), and asking stakeholders to brainstorm using what-if questions on possible design-related hazards. The returns from all stakeholders will then be compiled by the developer or his representative (e.g. a DfSP). For example, the design considerations during a GUIDE-2 review can include air-con ledges, a skylight at the atrium area, a building maintenance unit, etc. These considerations can lead to hazards such as open edges, lightning and falling objects. With the compiled list of design considerations and hazards, DfS review meetings can then be organised to focus on the issues identified.

It must be noted that the review should focus on hazards and controls not already covered in the Building Control Act and the Workplace Safety and Health Act. Since designers and contractors are already expected to comply with relevant regulations, the DfS review meetings should dedicate time to identify hazards peculiar to the structure being constructed. For example, if the construction site is next to a hospital, traffic to the hospital can be affected if the site entrance is placed along the road leading to the hospital. This can lead to emergency vehicles being delayed when trying to reach the hospital. In this case, even if having the entrance along the road to the hospital does not contravene any regulations, the entrance should be shifted. Compliance checks must still be conducted, but individual designers should be capable of doing that within existing building control processes. Nevertheless, it is important for developers and designers to be sensitive to WSH information, changes and incidents that may suggest a need to review specific designs even if they are deemed to be "standard" designs.

Another area of focus during DfS reviews is the interface between design considerations and construction, maintenance and operational activities. DfS review meetings should consider the activities that would be occurring during the relevant activities related to relevant design considerations and consider the possible hazards that can arise. For example, during precast construction, where key design considerations include the different types of precast components, the delivery of precast components will mean the presence of trailers and an increased number of lifting operations. These activities will increase the risk of being struck by moving vehicles and being hit by objects falling from height. This will mean that during GUIDE-2 or GUIDE-3 the design of the site layout should consider the shape, length and size of large precast components and ensure that site workers and delivery and lifting activities are segregated. These design changes should be tied to the different design considerations.

5.5 Safe Design Process in Australia

The Safe Design (SD) process in Australia is similar to the GUIDE process in Singapore, but it provides further details that are useful for DfS reviews. The SD process is described in the flowchart in Figure 5.4 and a summary of the process, as described in Safe Work Australia (2018), is provided below.

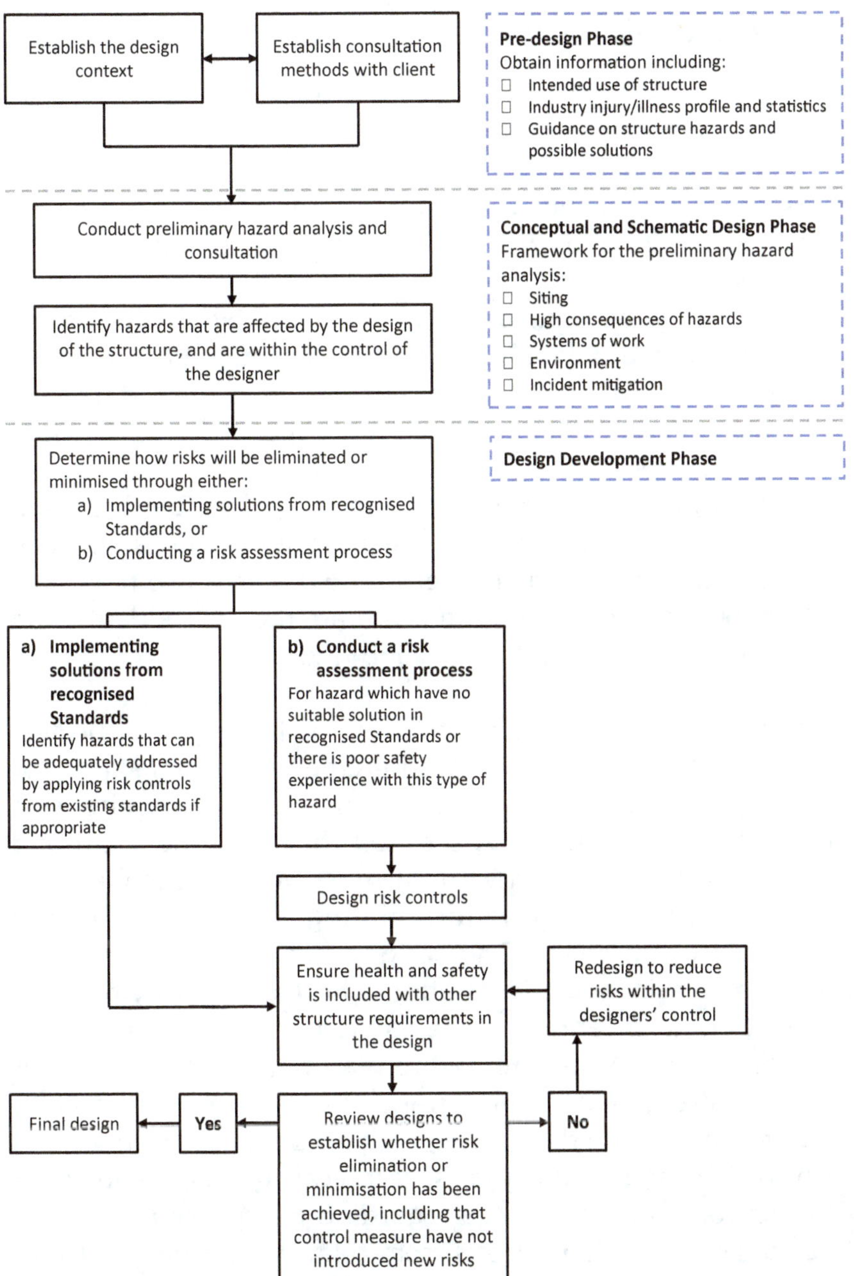

Figure 5.4   Safe design systematic approach (adapted from Figure 1 of Safe Work Australia (2018))

5.5.1 Pre-design stage

According to Safe Work Australia (2018), the pre-design phase of the SD process includes:

- Establishing the design context, focusing on the purpose of the structure, as well as the scope and complexity of the project.
- Establishing the risk management context by identifying the breadth of workplace hazards and relevant legislation, codes of practice and standards that need to be accounted for.
- Identifying the required design disciplines, skills and competencies.
- Identifying the roles and responsibilities of various parties in the project and establishing collaborative relationships with clients and others who influence the design outcome.
- Conducting consultations and research to aid in identifying hazards, assessing and controlling risks.

In this stage, the role of the client is to prepare a project brief that includes the safety requirements and objectives for the project which will ensure a shared and comprehensive understanding of safety expectations between the client and designer. Additionally, the client must provide the designer with all available information relating to the site that may affect safety and health. The designers on the other hand are required to ask their clients about the type of activities and tasks likely or intended to be carried out in the structure, including the tasks of those who maintain, repair, service or clean the structure as an important part of its use. Information required for the pre-design stage can be obtained from sources such as WSH and building laws, technical standards and codes of practice, industry statistics regarding injuries and incidents, hazard alerts or other reports from relevant statutory authorities, unions and employer associations, specialists, professional bodies representing designers and engineers, and research and testing done on similar designs.

5.5.2 Conceptual and schematic design stage

In this stage, preliminary hazard analysis should be conducted as early as possible. Firstly, the designers and other relevant stakeholders or duty holders involved in the preliminary hazard analysis should decide which hazards are 'in scope'. A hazard is 'in scope' if "it can be affected, introduced or increased by the design of the structure". At this stage, preliminary considerations should be given to possible ways to eliminate or substitute hazards.

Next, systems of work which are foreseeable as part of the construction method and the intended use of a structure as a workplace should also be identified in the preliminary hazard analysis. Likely or intended workflows, if known, will also be useful as part of the project brief prepared by the client. Furthermore, the project brief may also include any activities and systems with hazards specific to the nature of the structure. In the case of hazardous manual tasks, Australian WSH Regulations (WHS Regulations 61) specifies a designer must, as far as reasonably practicable, ensure that a structure is designed so as to eliminate the need to carry out any hazardous manual task in connection with it. Where this is not reasonably practicable, the designer must ensure the structure is designed so that the need for these hazardous manual tasks is minimised so far as is reasonably practicable.

Moreover, a systematic approach towards hazard identification should be done. The recommended framework in the Australian code of practice is elaborated in Table 5.1. It is noted that as in the case of the Singapore DfS guidelines, the use of checklists or hazard prompt lists is a useful preliminary hazard identification step, but SD or DfS review teams must not be limited by these lists.

Table 5.1 Framework for preliminary hazard identification (adapted from (Table 2 of Code of Practice for Safe Design of Structures (Safe Work Australia, 2018))

Source of hazard	Description
Siting of Structure	Potential design issues that may affect safety include: • proximity to adjacent property or nearby roads • surrounding land use • clearances required for construction equipment and techniques • demolition of existing assets • proximity to underground or overhead services — especially electricity lines • exposure of workers to adjacent traffic or other hazards • site conditions — including foundations, and construction over other assets or over water • safety of the public • use of adjacent streets
High consequence hazards	• storage and handling of dangerous goods • work with high energy hazards (e.g. pressure) • health hazards such as biological materials
Systems of work (involving the interaction of persons with the structure)	The systems of work (including cleaning and maintenance activities) that pose risks, for example: • rapid construction techniques, i.e. prefabrication versus *in situ* construction • materials to be used in construction • staging and coordination with other works • inadequate pedestrian or vehicle separation • restricted access for building and plant maintenance • hazardous manual tasks

Table 5.1   (*Continued*)

Source of hazard	Description
	<ul><li>working at height</li><li>exposure to occupational violence</li></ul> Consider both technical and human factors, including humans' ability to change behaviour to compensate for design changes. Anticipate misuse throughout the lifecycle.
Environmental conditions	<ul><li>impact of adverse natural events such as cyclones, floods, and earthquakes</li><li>inadequate ventilation or lighting</li><li>high background noise levels</li><li>facilities that do not meet workplace needs</li></ul>
Incident mitigation	A structure can increase the consequences of an incident if it has:<ul><li>inadequate entry and exit points</li><li>poorly situated assembly areas</li><li>inadequate access for emergency services</li></ul>

5.5.3 Design development phase

During the design development phase, the design team converts the design concepts into detailed drawings and technical specifications. The team will decide on design-related control measures, and prepare construction documentation. The completed design is then handed over to the client.

The design development phase should involve:

- Developing a set of design options corresponding to the hierarchy of control;
- Selecting the optimum solution. Balance the direct and indirect costs of implementing the design against the benefits derived;

- Testing, trialling or evaluating the design solution;
- Redesigning to control any residual risks, and
- Finalising the design, preparing the safety report and other risk control information needed for the structure's lifecycle.

When implementing solutions or control measures, legislative provisions, guidelines and standards have to be abided by. However, the mandatory regulatory requirements and standards may not adequately control WSH risks if applied to a situation outside that contemplated in the requirements or standard or if the requirement or standard is outdated. Therefore, the DfS review process must account for situations that may not be covered in the regulatory requirements and standards.

Moreover, RAs should be done to determine the severity of the risk, the effectiveness of existing control measures, the actions one should take to control the risks and the urgency of such actions to be taken. Risks are assessed by considering what could happen if someone is exposed to a hazard and the likelihood of it happening. In cases whereby the hazards and their associated risks are well known and there are well established and accepted effective control measures used in a particular industry that are suited to the circumstances of the workplace, then the formal assessment of the risk is not required and the controls can be implemented immediately. However, this author recommends that there must be effective monitoring processes to assure that the control measures continue to be effective throughout its implementation. The Australian code of practice recommended several RA methods for assessing design safety:

- fact-finding to determine existing controls, if any
- testing design assumptions to ensure that aspects of the design are not based on incorrect beliefs or anticipations on the part of the designer, for example as to how workers or others involved will act or react
- testing structures or components specified for use in the construction, end-use and maintenance phases

- consulting with key people who have the specialised knowledge and/or capacity to control or influence the design (for example the architect, client, construction manager, engineers, project managers and safety and health representatives); consulting directly with other experts (for example specialist engineers, manufacturers and product or systems designers) who have been involved with similar constructions, and
- when designing for the renovation or demolition of existing buildings, reviewing previous design documentation or information recorded about the design structure and any modifications undertaken to address safety concerns; and consulting professional industry or employee associations who may assist with RAs for the type of work and workplace.

Table 5.2 provides a framework to ensure all risks are addressed in the design and identifies the parties that should be involved in the various steps.

5.5.4 Reviewing control measures

As the design process continues, more detailed decisions are made. These decision points become opportunities for either eliminating or minimising risks. Designers should deliberately review design solutions at each decision point to establish the effectiveness of the risk controls previously selected and change them if necessary and so far as is reasonably practicable.

As indicated earlier, it is best if SD or DfS reviews involve the contractor and facilities management personnel with knowledge and experience in construction and maintenance processes. Their expertise will assist in identifying WSH issues overlooked during design.

The effectiveness of the control measures arising from the DfS review should be evaluated after construction. This ensures

Table 5.2   Design process (adapted from Table 3 of Code of Practice for Safe Design of Structures (Safe Work Australia 2018))

Step	Possible techniques	By whom
Identify solutions from regulations, codes of practice and recognised standards	• Consult with all relevant persons to determine which hazards can be addressed with recognised standards. • Plan the risk management process for other hazards.	Designer led. Client approves decisions.

Apply risk management techniques	• Further detailed information may be required on hazards, for example by:
	
	  ○ using checklists and referring to codes of practice and guidance materials, and
	  ○ job/task analysis techniques.
	
	• A variety of quantified and/or qualitative RA measures can be used to check the effectiveness of control measures.
	• Scale models and consultation with experienced industry personnel may be necessary to achieve innovative solutions to longstanding issues that have caused safety problems.
		Designer led. Client provides further information as agreed in the planned risk management process.

Discuss design options	• Take into account how design decisions influence risks when discussing control options.	Designer led. Client contributing.
Design finalisation	• Check that the evaluation of risk control measures in the design is complete and accurate.	Designer led. Client and designer agree

Table 5.2   (*Continued*)

Step	Possible techniques	By whom
	• Prepare information about risks to health and safety for the structure that remains after the design process.	on the final result.
Potential changes in construction stage	• Ensure that changes which affect design do not increase risks; for example the substitution of flooring materials could increase slip/fall potential and may introduce risks in cleaning work.	Construction team in consultation with designer and client.

identification of the most effective design practices and any design innovations that could be applied to future projects. The review can be carried out in a post construction workshop attended by all relevant parties involved in the project. On top of that, subsequent feedback from users to assist designers in improving their future designs may be provided through post occupancy evaluations, defects reports, accident investigation reports, information regarding modification, user difficulties and deviations from intended conditions of use.

5.6 Online Libraries Supporting Design for Safety Reviews

There are some online libraries supporting DfS Reviews and they include:

- Institution of Engineers, Singapore (IES) (2018) https://www.ies.org.sg/Publication/TechArticles
  - Design for Safety (DfS) Library: Examples of Hazards–Architectural Design

- o Design for Safety (DfS) Library: Examples of Hazards–Mechanical & Electrical Design

- Real Estate Developers' Association of Singapore (REDAS) (2019) http://www.redas.com/goodpracticeguide.html

  - o DfS and WSH good practice guide

- UK Design Best Practice–Promoting safety in design (n.d.) http://www.dbp.org.uk/

  - o Topics include Architecture, Civil, Structural, Mechanical, Electrical, Buildability, and Health & Wellbeing

- Occupational Safety and Health Council (2016) developed a series of worked DfS review examples in Hong Kong https://www.devb.gov.hk/

These online libraries are very useful, but the IES and Hong Kong resources are rather static and can be more comprehensive. The UK DBP database is relatively comprehensive, but more can be done to improve the search function and standardisation of the information in the database. As a whole, more can be done to improve these resources.

5.7 Challenges and Success Factors

According to Goh and Chua (2016), the top three perceived problems in practising DfS are client's cost concerns, contractors coming into the project too late, and inconsistency in DfS reviews or checks. To address these problems, the client needs to provide the necessary motivation to ensure that DfS is effectively implemented. As discussed earlier, ECI will be useful in improving DfS. Concurrent use of virtual design and construction (VDC) tools will help designers better foresee possible hazards.

Each DfS review will need to be effectively implemented. The following success factors for a DfS review are:

- Committed stakeholders who come to the DfS review meetings prepared and ready to contribute;
- An effective facilitator and coordinator for the DfS review process to ensure that all stakeholders are able to contribute to the review process;
- The design, WSH and operational competency of stakeholders (not just the facilitator) so that hazards, elimination and mitigation controls can be effectively identified;
- Information is available and reviewed prior to the meeting; and
- Detailed documentation and tracking of follow-up actions by stakeholders.

At the end, the personnel overseeing site operations (e.g. the Qualified Person for Supervision in the Building Control Act and facilities managers) will have to make sure that the DfS measures are effectively implemented on site and during operations of the structure.

From a systems thinking perspective, the successful implementation of DfS requires developers and designers to have a fundamental shift in their mental models. This is especially difficult because developers and designers may be uninterested about the WSH of workers downstream when compared with financial and design issues that are more pressing and that directly affect them. The main archetype that will be applicable to the successful implementation of DfS would be Eroding Goals and as advised in Chapter 3, it is important for the regulators, who are beyond market influences, to set the vision very clearly through regulations, standards and guidelines. This vision must be held consistently and constantly communicated with the industry until it becomes a shared vision. There must be open dialogue to listen to the challenges and confusion of implementing DfS, which are expected, and there must be resources provided to improve the competency of the industry in addressing design risks. At the same time, enforcement must be fair, strict and swift. These measures will help to close the performance gap between the desired and actual quality of DfS reviews in industry. The vision of WSH risks being eliminated and mitigated upstream should not waiver.

Review Questions

1. What are the roles of the different stakeholders in DfS?
2. Explain the GUIDE process and the differences between GUIDE 1–3.
3. Give examples of how design changes can improve: (i) WSH during construction, (ii) WSH during maintenance, (iii) WSH during use of facilities and (iv) WSH during demolition.
4. If you are the DfS Professional appointed by the developer, how would you ensure that a DfS review meeting is effective?
5. How can a developer ensure that DfS is successfully implemented in its projects?
6. Using the eroding goal archetype, describe how a DfS review process can deteriorate in terms of quality.

References

Behm, M. (2005). "Linking construction fatalities to the design for construction safety concept." *Safety Science*, 43(8), 589–611.

Bluff, L. (2003). "Regulating Safe Design and Planning of Construction Works — A review of strategies for regulating OHS in design and planning of buildings, structures and other construction projects." National Research Centre for OHS Regulation, Sydney.

Design Best Practice (n.d.), http://www.dbp.org.uk/ (accessed 11 March 2020).

Environment, Transport and Works Bureau (ETWB), Hong Kong Housing Authority (HKHA), and Occupational Safety and Health Council (OSHC) (2003). "Construction Design and Management — Worked Examples." http://www.devb.gov.hk/filemanager/en/content_29/cdm-worked%20 example.pdf. (accessed 6 July 2017).

Gambatese, J. A., Behm, M., and Rajendran, S. (2008). "Design's role in construction accident causality and prevention: Perspectives from an expert panel." *Safety Science*, 46(4), 675–691.

Goh, Y. M., and Chua, S. (2016). "Knowledge, attitude and practices for design for safety: A study on civil & structural engineers." *Accident Analysis & Prevention*, 93, 260–266.

Health and Safety Executive (2015). "The Construction (Design and Management) Regulations 2015." http://www.hse.gov.uk/construction/cdm/2015/ index.htm. (accessed 22 July 2015).

Institution of Engineers, Singapore (IES) (2018). "Technical Articles — Health and Safety Engineering" https://www.ies.org.sg/Publication/TechArticles. (accessed 11 March 2020).

López-Arquillos, A., Rubio-Romero, J. C., and Martinez-Aires, M. D. (2015). "Prevention through Design (PtD). The importance of the concept in Engineering and Architecture university courses." *Safety Science*, 73, 8–14.

Occupational Safety and Health Council (2016). "Worked Examples of Design for Safety." https://www.devb.gov.hk/filemanager/en/content_29/Design_for_Safety_Worked_Examples.pdf.   (accessed June 11, 2019).

Real Estate Developers' Association of Singapore (REDAS) (2019). "DfS & WSH good practice guide" http://www.redas.com/goodpracticeguide.html. (accessed 11 March 2020).

Safe Work Australia (2018). *Safe design of structures*, Safe Work Australia, Canberra.

Schulte, P. A., Rinehart, R., Okun, A., Geraci, C. L., and Heidel, D. S. (2008). "National prevention through design (PtD) initiative." *Journal of Safety Research*, 39(2), 115–121.

Toh, Y. Z., Goh, Y. M., and Guo, B. H. W. (2016). "Knowledge, Attitude and Practice of Design for Safety: A Study on Multiple Stakeholders in the Construction Industry." *J. Constr. Eng. and Manage. — Am. Soc. of Civ. Eng.*, 143(5), 04016131.

WorkCover New South Wales (2001). "CHAIR — Safety in design tool." WorkCover New South Wales, Sydney.

Workplace Safety and Health Council (2015). "Design for Safety." https:// www.wshc.gov.sg/. (12 Oct 2015).

Workplace Safety and Health Council (2016). "Workplace Safety and Health Guidelines — Design for Safety." https://www.wshc.sg/files/wshc/upload/ cms/file/WSH_Guidelines_Design_for_Safety(1).pdf. (6 July 2017).

Appendix

Project Title:　　　　　Company　　　　　Review date:　　　　　Next review date:

Conducted by:

Process/ Location:

S/No.	Design consid- eration	Hazard identi- fied	Severity	Likelihood	Risk level	Design out?	Proposed control	Severity	Likelihood	Risk level	Further review required	Action by (date)

CHAPTER 6

Overview of Workplace Safety and Health Management Systems

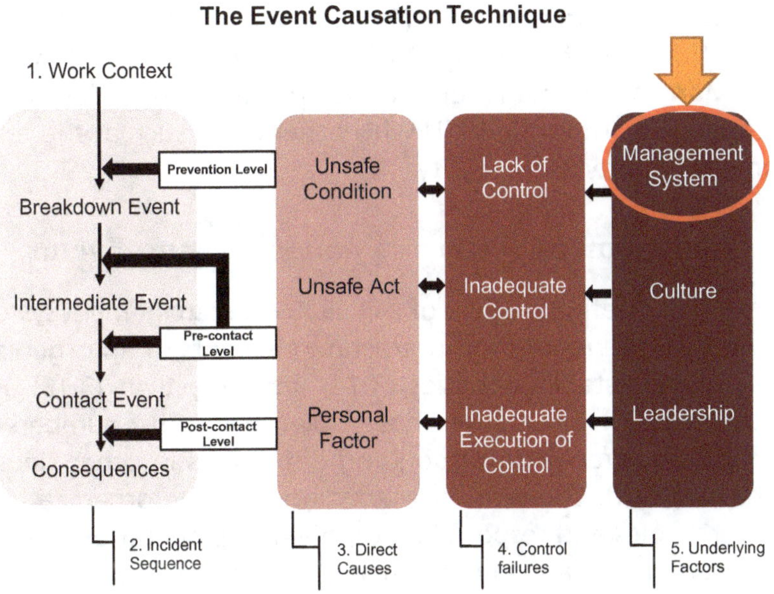

As identified in the Event Causation Technique (ECT) diagram above, the workplace safety and health (WSH) management system is an important underlying factor that can ensure that suitable risk controls are implemented effectively to prevent incidents.

6.1 Introduction

This chapter provides an introduction to workplace safety and health (WSH) management systems. ISO 45001:2018 (ISO, 2018) will be used as an example of a WSH management system framework. It must be noted that there are other WSH management system frameworks, for example, CP 79: 1999 (Singapore Standards Council, 1999), ANSI/ASSE Z10-2012 (R2017) (American National Standard (ANSI)/American Society of Safety Engineers (ASSE), 2017) and the International Labor Organization (ILO) Guidelines on occupational safety and health management systems (International Labor Organization, 2001). There are also numerous WSH management system frameworks developed by different companies such as ExxonMobil's Operations Integrity Management System (OIMS) (ExxonMobil, n.d.), and DNV GL's International Sustainability Rating System™ (ISRS) (DNV GL, n.d.). Many of the management system frameworks developed by companies integrate WSH with other areas such as the environment, quality and security.

6.2 Management System as an Underlying Factor

According to definition 3.10 of ISO 45001, a management system is a "set of interrelated or interacting elements of an organization (3.1) to establish policies (3.14) and objectives (3.16) and processes (3.25) to achieve those objectives". The numbers in parentheses refer to corresponding definitions in other clauses in the standard. The term *elements* include "the organization's structure, roles and responsibilities, planning, operation, performance evaluation and improvement." Based on this definition of a management system, an occupational health and safety (OHS)[1] management system is defined as a "management system or part of a management system used to achieve OHS policy", while OHS policy refers to "intentions and direction of an organization (3.1), as formally expressed by its top management (3.12)".

As can be seen, a management system is essentially a set of fundamental management processes and structures. The overall purpose of a management system is to help an organisation

---

[1] ISO 45001:2018 uses OH&S or OHS, which has the same meaning as WSH. The terms will be used interchangeably.

set and achieve its policies and goals, and continually improve itself. The basic Plan–Do–Check–Act (PDCA) cycle is fundamental to a management system and one of the key purpose of the different elements is to help managers execute and maintain the PDCA cycle.

It is important to understand the difference between a management system and a specific control measure. A control can be a barricade, a training course, a set of safety rules or personal protective equipment (PPE). This will typically manage the direct causes (e.g. unsafe acts, unsafe conditions and personal factors) and these are usually event level remedies (see Chapters 1 and 3) that stand in contrast to a management system intervention, which focuses on the more fundamental elements of a management system, as highlighted earlier. Control measures are usually identified through Risk Assessments (RAs) and they are meant to reduce the risks posed by a hazard. In contrast, a management system influences the patterns and trends of unsafe acts and conditions in the workplace.

Let us take, for example, a worker working unsafely at height on wooden planks that are too flexible and unstable. An event level answer to the problem would identify the planks as unsafe, and then change them to ensure that the worker is safe. However, a more systemic approach would be to identify that the planks are unsafe, then determine if the same type of planks are widely used in the organisation and whether workers are able to identify the danger that the planks pose to workers working on them. If the planks are used on all the sites in the organisation and workers are not able to perceive the planks as hazards, the systems approach identifies that there is a dangerous pattern indicating that management system elements, for example, the RA process (ISO 45001:2018 Clause 6.1.2), procurement process (Clause 8.1.4), and evaluation of training effectiveness (Clause 7.2), may be inadequate. Using the procurement process as an example, it may be discovered that the procurement process did not include WSH criteria and feedback from site personnel were not included prior to the procurement of material. The procurement department may have been overly focused on price, and may not have been able to account for WSH criteria. Experienced site personnel should be required to provide input prior to purchase. The procurement process can then be improved by including a requirement for experienced site personnel to feedback on purchase specifications. A feedback form can also be developed, and relevant site personnel should be identified as relevant subject matter experts for different types of purchase.

The above example shows that by improving the management system elements, in this case the procurement procedure, the organisation as a whole becomes safer — not only will the planks used by workers on site be safer, all newly purchased equipment should be safer due to input from experienced site personnel. Thus, with reference to the Event Causation Technique (ECT), a management system is an important underlying factor that interacts with culture and leadership to influence the effectiveness of all WSH controls and direct causes that can lead to incidents.

6.3 Overview of ISO 45001:2018

The following will provide an overview of ISO 45001:2018 as an illustration of the key elements of a WSH management system. Readers are encouraged to read the actual standard for more details.

Figure 6.1 shows the management system framework of ISO 45001:2018. This framework is aligned with other ISO management system standards such as ISO 9001 (quality) and ISO 14001 (environmental management). Each of the elements in the framework will be discussed in the following sections. It is

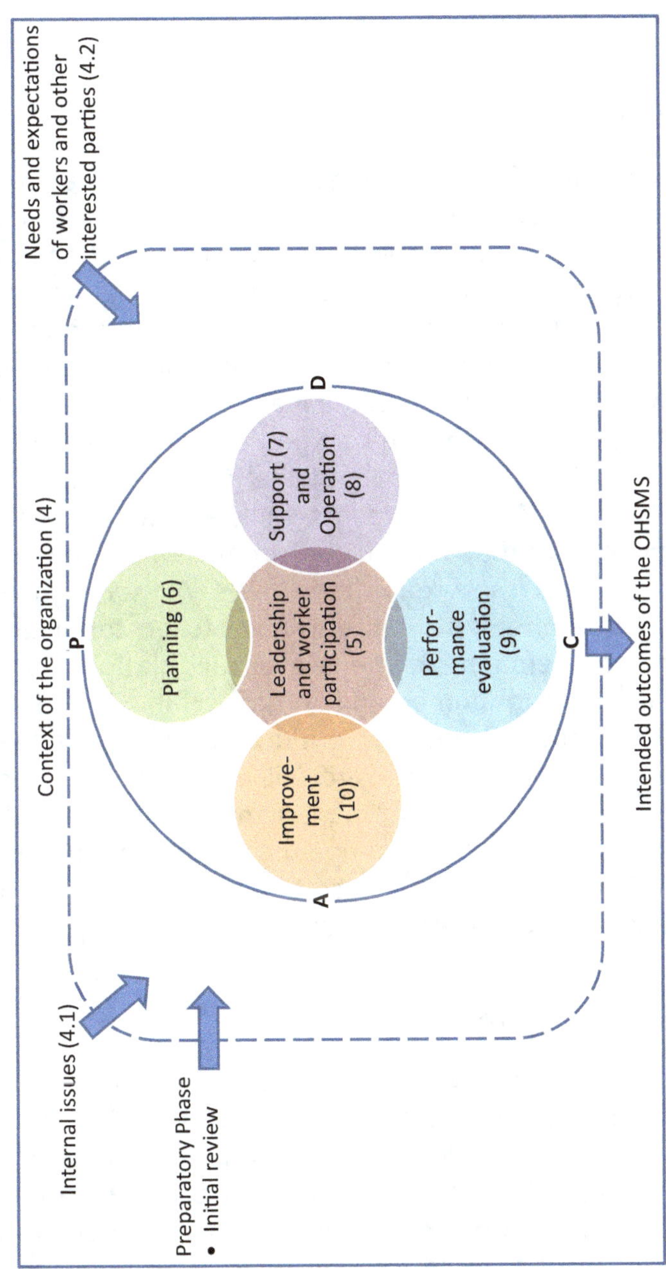

Figure 6.1   ISO 45001:2018 Management system framework; adapted from ISO (2018)

noted that the "initial review" element was inserted into the recommended framework by the author.

6.3.1 Initial review

With or without a formal management system, organisations have their own way of managing themselves. These processes may or may not be documented, but they are happening. When an organisation is trying to set up a formal and documented management system, they will have to first conduct an initial review, which compares existing management processes with the processes stipulated in the management system standard that they are trying to be certified to or model after, e.g. ISO 45001 or CP79:1999. As part of the initial review, existing risk management (RM) documents will have to be reviewed and additional RAs may have to be conducted to understand the WSH risks that the organisation faces. A thorough check against legal requirements will also have to be conducted to understand the demands of the relevant legislations. Existing WSH trainings, programmes and initiatives should be documented. The resources used to conduct existing management system processes, programmes, etc. will also have to be clearly documented. Once the current management approach is documented, it should be compared against the WSH management system standard to identify gaps, so that specific actions can be identified and implemented to close the gaps.

6.3.2 Success factors

The ISO 45001:2018 Clause 0.3 highlighted a non-comprehensive set of success factors for an effective WSH management system. The 11 factors are classified into five categories as shown in Table 6.1. First, for a WSH management system to be successful, it is important for the top management to provide WSH leadership, build strong safety culture and provide sufficient resources for WSH management. Second, the top

Table 6.1 Categorisation of success factors identified in ISO 45001:2018

Category	Success factors identified in ISO 45001:2018
(I) Top management, culture and resource allocation	a) top management leadership, commitment, responsibilities and accountability;
	b) top management developing, leading and promoting a culture in the organisation that supports the intended outcomes of the OHS management system;
	e) allocation of the necessary resources to maintain it;
(II) Policies, objectives and compliance	f) OHS policies, which are compatible with the overall strategic objectives and direction of the organisation;
	j) OHS objectives that align with the OHS policy and take into account the organisation's hazards, OHS risks and OHS opportunities;
	k) compliance with its legal requirements and other requirements.
(III) Communication	c) communication;
	d) consultation and participation of workers, and, where they exist, workers' representatives;
(IV) Effective and integrated processes	g) effective process(es) for identifying hazards, controlling OHS risks and taking advantage of OHS opportunities;
	i) integration of the OHS management system into the organisation's business processes;
(V) Continual improvement	h) continual performance evaluation and monitoring of the OHS management system to improve OHS performance;

management must lay down clear WSH policies and objectives that guide the implementation of the WSH management system and ensure compliance to legal and other requirements. Third, the policies and objectives must be clearly communicated with employees and stakeholders, and there must be opportunities for consultation and participation on WSH matters. Fourth, the WSH management processes, especially the RA process, must be effective and integrated into the organisation's business processes. Last, there must be continual improvement processes to improve WSH performance.

As can be observed, the key success factors are aligned with the ECT, which highlighted the importance of leadership, culture and management system. This set of 11 success factors is a useful reminder to managers of what to focus on when managing a WSH management system.

6.3.3 Context

A WSH management system does not exist in a vacuum. The context that it exists in contains different issues that will affect the way that the WSH management system should be designed, operated and maintained. These issues can be classified into internal and external issues. BS 45002:2018 (BSI Standards Limited, 2018) provides further guidelines on what some of the possible internal and external issues are. External issues include relationships with external providers such as contractors or suppliers, new technologies, cultural, social and political factors, and legislations. Internal issues include the size, nature and activities of the organisation, the way the organisation is managed and its business objectives, resources, knowledge and competence, and planned or foreseeable changes and how these are managed.

When identifying relevant issues in the context, BS 45002:2018 (BSI Standards Limited, 2018) recommended approaches such as "what-if" questions for organisations that run less complex operations and are smaller in size. For other organisations, structured methods such as SWOT (Strengths, Weaknesses, Opportunities and Threats) or PESTLE (Political, Economic, Social, Technological, Legal, Environmental) analysis can be used.

One key internal context that must be clearly established are the needs and expectations of workers and other interested parties. Thus, Clause 4.2 of ISO 45001:2018 requires the organisation to identify workers ("workers" include managerial, supervisory and non-managerial staff) and other interested parties that are relevant to the WSH management system, their needs and expectations, and the relationships between these needs and expectations and relevant legal requirements.

Clause 4.3 specifies that the scope of the WSH management must be determined. The scope, i.e. boundaries and applicability, can be determined based on the types of activities, locations, products and services that the organisation has. For example, an organisation may choose to exclude certain locations from the WSH management system because the location is regulated by some prescriptive legal requirements that make the processes in the WSH management system unsuitable.

6.3.4 Leadership and worker participation

6.3.4.1 Leadership and commitment (Clause 5.1)

Unlike its predecessor, OHSAS 18001:2009, ISO 45001:2018 places additional attention on leadership and worker participation. Clause 5.1 stipulates 13 sub-clauses on the top management's responsibilities. These include taking overall responsibility and accountability for WSH, ensuring the establishment of the WSH management system, integrating the WSH management system into the organisation's business processes, ensuring adequate resources, communicating the importance of effective WSH management and conformance, etc. The responsibilities are essentially elaborations of the 11 success factors highlighted earlier as well as top management's responsibilities over processes highlighted in other parts of ISO 45001:2018.

6.3.4.2 WSH policy (Clause 5.2)

WSH policy is one of the most fundamental elements of a WSH management system. It is essentially a statement that top management issues to demonstrate and communicate their commitment to WSH. However, despite its theoretical importance, the policy statement does not have a direct impact on an organisation's WSH, and it is becoming too easy for organisations to simply employ consultants to help them draft suitable policy statements. The most important part of Clause 5.2 is perhaps the phrase "[to] establish, implement and maintain" — especially the word "implement". It is important for the policy, which is the key guidance for the rest of the management system, to be implemented. To be implemented means that the management system does not just exist on paper but is actually being used by people on the ground. "Maintained" means it is being utilised, checked and improved in a sustainable manner. This requires active effort on the part of the organisation. Many systems start well, but deteriorate due to lack of maintenance. Many of the elements of ISO 45001:2018 (such as performance evaluation and improvement) are designed to ensure the active maintenance of the system.

6.3.4.3 Organisational roles, responsibilities and authorities (Clause 5.3)

The top management is required to assign responsibilities and authorities for relevant roles in the WSH management system. This can include roles such as a management representative overseeing the WSH management system, RA teams, WSH officers, WSH committee members and WSH internal auditors. The roles, responsibilities and authorities must be documented and communicated to all interested parties. Workers (including managers and supervisors) shall assume responsibility for those aspects of the WSH management system over which they have control. However, ISO 45001:2018 clearly states that top management is still accountable for the functioning of the WSH management system.

One of the most basic ways to assign roles, responsibilities and authorities is to have an organisation chart and accompanying descriptions of the relevant WSH roles, responsibilities and authorities for each position in the organisation chart. In addition, for each process or project, the roles and responsibilities can be described based on "RACI", which stands for Responsible, Accountable, Consulted and Informed. "Responsible" refers to managers or persons-in-charge of the actual implementation of the work or process. "Accountable" refers to the owner or champion for the work or process who signs-off or approves the work or process. The person accountable will take overall responsibility for the work or process. "Consulted" refers to interested parties that need to give inputs to the work or process to enable effective implementation. And "Informed" refers to people that need to be kept in the picture, but do not need to be formally consulted and that are not directly involved in the work or process.

6.3.4.4 Consultation and participation of workers (Clause 5.4)

ISO 45001:2018 requires a process(es) for the consultation and participation of workers (and their representatives) at all levels and functions. Participation is defined as involvement in decision-making, while consultation is defined as seeking views before making a decision. The consultation and participation mechanisms must be supported by the necessary resources, training and time. In addition, management must determine and remove or minimise obstacles or barriers to participation. Some of these barriers and obstacles include failure to respond to worker inputs or suggestions, language or literacy barriers, reprisals or threats of reprisals, and policies or practices that discourage or penalise worker participation.

ISO 45001:2018 emphasises the consultation of non-managerial workers on issues such as WSH needs and expectations, WSH policy, assignment of roles and responsibilities, WSH objectives and plans, controls for outsourcing, procurement and contractors, WSH monitoring, measurement and evaluation, audit and continual improvement. In terms of the participation of non-managerial workers, the following are emphasised: mechanisms for consultation and participation, RA and controls, competency issues, communication processes, incident investigation and corrective actions.

6.3.5 Planning (Clause 6)

The organisation must create plans to achieve the policy and the desired outcomes of the WSH management system, i.e. prevention of incidents and reduction of WSH risks. The planning element consists of actions to address WSH risks, opportunities (Clause 6.1) and objectives, and planning to undertake them (Clause 6.2). Clause 6.1 is focused on the RA process, which is similar to the process covered in Chapter 4 on RM. The management system will have to specify the procedure for the RM process. Organisations will have to take reference from relevant national standards or guidelines on RA and RM. However, there could be differences between the terminology used in ISO 45001:2018 and national standards or guidelines. For example, in the Singapore Risk Management Code of Practice (RMCP) (see Chapter 4), RA refers to hazard identification, risk evaluation and risk control, but in ISO 45001:2018, RA is not clearly defined, and hazard identification and assessment of risks are separately defined. Putting the differences in terminology aside, both emphasise the importance of conducting a systematic RA (hazard identification, risk evaluation and risk control).

It must be noted that ISO 45001:2018 (Clause 6.1.2.3) highlights the need to have process(es) in place to assess WSH opportunities (in contrast to hazards and risks) to enhance WSH performance. The standard also requires considerations such as design of work areas, processes, installations, machinery/equipment, operating procedures and work organisation, which is aligned with the concept of Design for Safety (see Chapter 5).

Clause 6.1.3 highlights the importance of having a procedure for identifying and accessing the legal and other Safety, Health and Environmental (SHE) [management] requirements that are applicable to the organisation. This may seem straightforward, but due to the wide range of applicable legislations and standards, it is actually a difficult task to keep updated on the changes and evaluate the possible impact of these legislations and standards. Thus, many organisations subscribe to information services to keep themselves updated.

Clause 6.1.4 requires organisations to plan for actions to address risks, opportunities, and legal and other requirements, as well as to prepare for and respond to emergency situations (Clause 8.2). The plan must also integrate and implement the actions into relevant processes and evaluate the effectiveness of the actions.

Clause 6.2 is concerned with objectives. Policies identify issues of great importance to the organisation; based on the policy statement and WSH plans, objectives can then be created. For instance, if the organisation is concerned with the exposure to chlorinated solvents, then they can set an objective to eliminate the use of chlorinated solvents, which are used in degreasing and cleaning activities but are also carcinogenic, in all manufacturing lines by December 2021. The objective can also be supported by more specific targets, e.g. to replace 50% of chlorinated solvents in all manufacturing lines by June 2021 and to replace 75% of chlorinated solvents by September 2021.

Once a set of objectives and targets has been determined, the organisation can have a plan, which identifies:

(a) what will be done;
(b) what resources will be required;
(c) who will be responsible;
(d) when it will be completed;
(e) how the results will be evaluated, including indicators for monitoring;
(f) how the actions to achieve OHS objectives will be integrated into the organisation's business processes.

A series of activities, campaigns and events that help the organisation achieve its policies, objectives and targets will be contained in the plan. The plan must have a clearly identified timeline and allocation of responsibilities and accountabilities.

6.3.6 Support (Clause 7)

To support the functioning of the different elements of the WSH management system, ISO 45001:2018 identified five areas of support: resources, competence, awareness, communication and documented information. Resources are relatively self-explanatory and have been highlighted in some of the earlier clauses. Competence is a critical area in WSH and workers (including managers and supervisors) need to be trained so that they can perform their work safely. Competence goes beyond mandatory training required by the government. Mandatory training often only provides the basics, and on-going customised training is more critical. Thus, a training needs analysis must be conducted based on the RAs, applicable legislations, WSH standards, competency standards and other relevant documents. Besides training, experience is also critical in ensuring competence.

Besides competence, workers need to be aware of a range of WSH management system information, such as the policy and objectives, benefits of improved WSH performance, consequences of poor WSH, incidents and investigation outcomes, RA outcomes and actions, and their right to stop work when the WSH risk is not acceptable.

Communication refers to the communication processes needed for internal and external WSH communications. The processes must include determination of:

(a) What needs to be communicated;
(b) When to communicate;
(c) With whom to communicate (internal, contractors, and interested parties); and
(d) How to communicate.

The communication processes must take into account possible communication barriers like literacy levels, language and cultural diversity, and disabilities.

Documentation is a fundamental process in a formal management system, but the level of documentation must be proportional to the risk level, level of complexity and size of the operations. Legal requirements are also a major factor to consider when considering documentation. For instance, in Singapore and many countries, all RAs must be documented. At the same time, the control of documented information needs to be robust to ensure the accessibility and reliability of the information. A common problem in organisations is the failure to ensure the proper versioning of documents, leading to outdated documents being used at the workplace. Processes for document control are critical to prevent miscommunication and the application of out-dated procedures.

6.3.7 Operation (Clause 8)

Clause 8 covers the processes that are needed to prevent WSH incidents and this relates to the plans and actions identified during planning. Operational planning and control covers elements such as determining necessary process criteria that define safe operations, safety work procedures, and work instructions. The controls are identified based on the RA, and those that are similar across different RAs can be grouped together as general control measures, e.g. basic WSH rules, hazardous chemical procedures, programmes for the maintenance and repair of facilities, housekeeping procedures, traffic management plans, occupational health programmes, permit-to-work systems and PPE programmes. High risk activities that present specific hazards will require a dedicated set of controls and procedures for the activities based on the hierarchy of control discussed in the earlier chapter on RM.

An important process during operation is management of change (MOC) (Clause 8.1.3). MOC is implemented whenever there are changes in the workplace that influence WSH. These changes can be temporary (e.g. a truck breaking down at a construction site during a casting operation) or permanent (e.g. a new pressure vessel on the factory floor). The types of changes highlighted in ISO45001:2018 include:

- New products, services and processes, or changes to existing products, services and processes, including workplace locations and surroundings, work organisation, working conditions, equipment and workforce.
- Changes to legal and other requirements.
- Changes in knowledge or information about hazards and risks.
- Development in knowledge and technology.

The organisation needs to define processes to evaluate these changes, assess the WSH risks and opportunities, and implement controls to manage the risks and opportunities. It is important for organisations to consider possible unintended consequences of changes. One example is when personal mobility devices (PMDs) were introduced to cut down walking time in a workplace, it unintentionally led to increase in PMD collision accidents and fire incidents during charging of the PMDs. Thus, relevant existing RAs have to be identified and reviewed.

Procurement (Clause 8.1.4) is focused on the establishment of WSH requirements for products and services to be purchased. The purchasing department will have to identify the standards that purchases have to comply with, write up relevant WSH specifications such as required certifications, expectations for the WSH performance of the service provider and required audits on the service provider or supplier. The procurement department will have to inform the supplier or service provider on what is expected of them in terms of their WSH management systems. The purchasing organisation may expect the supplier to be involved in the WSH activities of the organisation, e.g. participate in safety campaigns, have a similar level of training and similar RA. Pre-approvals of requirements, specifications and procedures for the purchase of chemicals, machinery, and equipment will require the involvement of relevant experts and a clear process for checks prior to purchase. The WSH performance of suppliers and the goods that they have supplied should also be recorded and evaluated. This information can then be used for future purchase evaluation. Goods, equipment and services should be inspected and verified in terms of the WSH specifications and documented for future purchases.

Another important area is control over contractors. It is common for organisations to use contractors in their work. There are advantages in using contractors with the expertise that the organisation lacks, but organisations need to manage the contractors carefully. Contractors can be unfamiliar with the work environment, WSH requirements and hazards of the client organisation, and they may have a weak WSH management system. Thus, it is important to establish clear criteria to select contractors who can demonstrate their capability to manage WSH. To do this, many clients rely on WSH certifications to standards like ISO 45001:2018 and other safety schemes or standards. In Singapore, experience shows that relying only on certifications and other schemes like bizSAFE[2] may not be the best gauge of good WSH performance. A mix of information like interviews, audit reports, client's own experience with the contractors, and references from other clients are critical information in the contractor selection process. Accident and incident records are useful sources of information, but many companies are simply lucky and serious accidents may be too rare to provide a good indication. Nevertheless, accident and incident records must still be reviewed when selecting contractors. Contractors, especially term contractors or those dealing with larger projects, will have to be evaluated and reviewed for their WSH performance and diligence in implementing their WSH management systems. Some basic checks would include on-going inspections, review of method statements and RAs, and review of safety and health committee meeting minutes. An important aspect is to determine how senior management and line management view WSH. If they see it as the job of the WSH department, then it is an indication that WSH is not taken seriously.

Suppliers and contractors are critical stakeholders in an organisation's management of WSH performance. Contractors frequently deal with highly hazardous processes. Due to the short-term nature of work, varying levels of risk perception and other factors, contractors can cause WSH incidents. Clients, through their contracts and WSH department, will have to monitor WSH management activities of the contrac-

---

[2] A scheme under the Workplace Safety and Health Council, which is described as "a five- step programme that assists companies to build up their WSH capabilities so that they can achieve quantum improvements in safety and health standards at the workplace."

tors from selection to evaluation and monitoring. It is critical to store organisational knowledge of different contractors and suppliers so that poor performing contractors and suppliers will not be re-engaged. It is important to take a partnering approach, where the client helps the contractors and suppliers improve and use incentives and recognitions to motivate the contractors and suppliers to progress in terms of WSH management.

After the contractor has been selected, there must be frequent meetings to monitor how the contractor is progressing in its work and whether critical WSH procedures are implemented, e.g. RA, inspections, WSH meetings, incident investigation and audits. The contractors' resources such as plants, machinery, equipment and chemicals must be checked for their compliance to the company's requirements before they can be brought into the workplace. There can also be the sharing of incident learning and good practices among the contractors or sub-contractors.

Outsourcing has similarities to the engagement of a contractor. The key difference is that the function or process being outsourced can actually be performed in-house, but the organisation chose to focus on their core business and outsource non-core functions or processes. The key reasons for outsourcing are usually cost, resources and lack of time. As in the case of engaging contractors, the outsourced companies must be carefully selected and monitored.

Organisations also need to identify and prepare for the emergencies that can occur in their workplaces. The processes to prepare for and respond to these emergencies need to be established, implemented and maintained. These processes include establishing an emergency response plan (including first aid), training for relevant personnel, periodic tests and exercises, performance evaluations, and communication of the plan to workers and interested parties.

6.3.8 Performance evaluation (Clause 9)

The key elements under performance evaluation include monitoring, measurement, analysis and performance evaluation (Clause 9.1), and internal audit (Clause 9.2). Performance evaluation processes cover both quantitative and qualitative measures of WSH performance. These will be closely related to the policies, objectives and targets. At the same time, the measures will also be guided by the RA, especially the high-risk hazards and critical controls. The measures can also be split into proactive and reactive measures. Proactive measures monitor the implementation of controls and management activities, which are not dependent on the occurrence of incidents and accidents, i.e. reactive measures. Some of the proactive measures include inspections, audits, evaluation of training effectiveness, behaviour-based observations, and perception or safety culture/ climate surveys. Reactive measures count the number and rate of incidents and accidents, e.g. accident frequency rate, accident severity rate and fatality rate. Evaluation of compliance is directly related to the legal and other WSH requirements identified during planning. During implementation and operations, there should be periodic audits and assessments to determine if the organisation is complying with all applicable legislations and standards.

An audit is an important activity in the evaluation of compliance to the management system. This is unlike inspection, which is usually only focused on event level issues. According to ISO 45001:2018, an audit is a "systematic, independent and documented process for obtaining 'audit evidence' and evaluating it objectively to determine the extent to which 'audit criteria' are fulfilled". There is a strong emphasis on evidence and the evaluation needs to be thorough. An audit is usually based on the review of documents, interviews with employees and physical inspections. The three processes produce evidence for triangulation so that the auditor can form a credible impression of how well the management system is established and maintained. The audits can be conducted internally or by an external auditor. In the Singapore construction industry, for example, the Building and Construction Authority (BCA) and the Ministry of Manpower (MOM) both require safety and health audits on construction contractors. The BCA requires ISO 45001:2018 or OHSAS 18001 (predecessor of ISO 45001:2018), and ISO 14000 and ISO 9000 certification for contractors with a rating of B2 and above. The contractor ratings determine the size of government contracts that the contractors can bid for. Certification audits are usually conducted by independent and external auditors. These contractors need to be audited annually. In contrast, MOM requires six monthly independent and external audits for projects that are worth S$30 million and above, while projects with a contract sum of less than S$30 million are required to conduct internal reviews of the management system every six months. Past experience has shown that audits can be compromised due to commercial pressure on the auditors. Auditors can be pressurised by their paymaster, the contractors, to be more lenient. Thus, the professionalism of the auditors and the WSH culture of the contractors play an important role in the success of the audit regimes.

"Management review" refers to a high-level review of how effectively the management system is performing, which allows top management to make necessary changes to system elements like policies, resources and processes, to improve the performance. There is a wide range of documents that the management can review to get a sense of how established, implemented and well-maintained the WSH management system is. These documents would include incident investigation reports, audit reports and performance indicators derived from objectives and targets.

6.3.9 Improvement (Clause 10)

Incident investigation is one of the most fundamental management processes and the organisation must respond strongly and seriously to any incident that occurs. This is covered in more detail in Chapter 3.

Besides incidents, nonconformity can be uncovered during operations or inspections, such that employees and contractors may be found to violate WSH rules or procedures. Based on investigations into the incidents or nonconformity, corrective actions will be proposed; these should be recorded and their effectiveness and implementation tracked. Corrective actions are focused on removing causes of nonconformity and preventing their recurrence. On the other hand, opportunities for continual improvement can also be identified during the investigation or operations. These opportunities are not linked to any specific nonconformity or incidents, but they do enhance WSH performance.

6.4 Management Systems as Balancing Loops

The PDCA cycle used in ISO 45001:2018 is also known as the Deming cycle. The basic premise of the PDCA cycle is the balancing loop as discussed in Chapter 2, which is contextualised in Figure 6.2. During planning, there are different processes to facilitate the determination of WSH policies, which are translated into measurable objectives and targets. At the same time, WSH controls are planned ("Plan") for and then implemented during operations ("Do"). Checks are then conducted ("Check") to measure the WSH performance and level of conformity to planned controls. The performance gap determined during checking becomes the basis for improvement actions ("Act") determined by management. However, there is a delay between improvement actions and measured performance because audits, inspections and checks are conducted periodically and are not comprehensive.

Figure 6.2 fits the systems archetype of "Balancing a process with delay" (see Table 3.4 of Chapter 3) perfectly. Imagine a

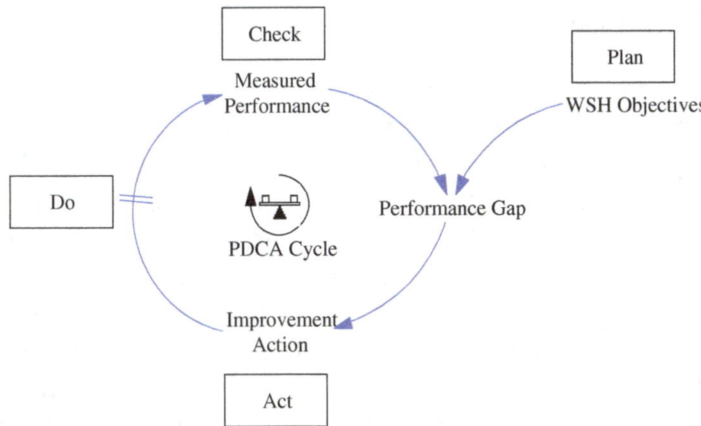

Figure 6.2   PDCA cycle as a balancing loop with delay

person trying to achieve a desired temperature of the shower water by adjusting the amount of hot water, but there is a delay between his action and the change in temperature. To make things worse, the duration of the delay is unknown. The longer the delay between his action and the change in temperature, the more difficult it is to achieve the desired temperature. The person will either be having very cold water or very hot water. Similarly, to achieve a WSH objective, the management might put in too much resources or effort, possibly leading to impressive short-term WSH performance, but at the sacrifice of other organisational objectives. Or they might put in too little resources or effort, resulting in poor WSH performance.

Based on the advice in Table 3.4 of Chapter 3, the management needs to exercise patience, build an information system that gives quicker feedback and actively seek the inputs of frontline staff. These are sound advice that are aligned with the guidance provided in ISO 45001:2018. Since WSH management systems are essentially control systems, other systems thinking archetypes that are especially relevant are "Eroding or drifting goals", "Fixes that fail" and "Shifting the burden (addiction)".

6.5 Conclusions

The WSH management system manages the risk of WSH incidents. It is important for managers to think systemically and design processes and structures that ensure that suitable and effective controls are identified, implemented and maintained. This will then prevent the unsafe acts and conditions that cause incidents to happen. ISO 45001:2018 is a comprehensive document that provides a useful illustration of the key components expected of a robust and reliable WSH management system. From a systems thinking perspective, management systems are essentially balancing loops constructed to help organisations achieve a desired level of WSH performance. The guidance provided by the relevant system archetypes are useful advice for managers trying to improve WSH performance.

Review Questions

1. What is the difference between a risk control and an element or sub-system of a management system?
2. Give three examples of WSH Policies, Objectives and Targets for a large food and beverage outlet.
3. Explain how contractors and suppliers should be managed in terms of WSH.
4. What are the potential problems with audits and how can they be mitigated?
5. Explain what management review is and why it is important.
6. Describe how the systems archetypes of "Eroding or drifting goals", "Fixes that fail" and "Shifting the burden (addiction)" can be used to guide the WSH management efforts for a transport operator.

References

American National Standard (ANSI)/American Society of Safety Engineers (ASSE). (2017). *ANSI/ASSE Z10-2012 (R2017) Occupational Health and Safety Management Systems*.

BSI Standards Limited. (2018). *BS 45002-0:2018 Occupational health and safety management systems — Part 0: General guidelines for the application of ISO 45001*.

DNV GL. (n.d.). ISRSTM: for the health of your business — Oil & Gas — DNV GL. https://www.dnvgl.com/services/isrs-for-the-health-of-your-business-2458. (accessed April 3, 2018).

ExxonMobil. (n.d.). OIMS: A disciplined management framework | Exxon- Mobil. http://corporate.exxonmobil.com/en/company/about-us/safety-and-health/operations-integrity-management-system. (accessed April 3, 2018).

International Labor Organization (ILO). (2001). Guidelines on occupational safety and health management systems. http://www.ilo.org/wcmsp5/groups/public/---dgreports/---dcomm/---publ/documents/publication/wcms_%20publ_9221116344_en.pdf. (accessed September 2, 2020).

ISO. (2018). *ISO 45001:2018 Occupational health and safety management systems*.

ISO 45001. (2018). Professional Safety, 63(2), 53–55. http://libproxy1.nus.edu.sg/login?url=https://search-proquest-com.libproxy1.nus.edu.sg/docview/1993301613?accountid=13876 (accessed September 2, 2020).

Singapore Standards Council. (1999). *Code of practice for safety management system for construction worksites*.

CHAPTER 7

Safety Culture and Leadership

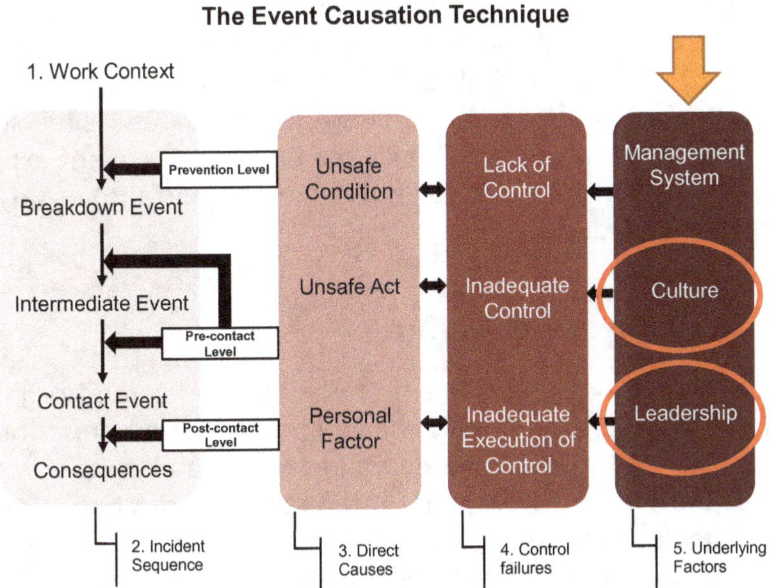

Safety culture and leadership are underlying factors within the Event Causation Technique (ECT) framework (see ECT diagram above). They influence the effectiveness of the management system and the risk controls. Unlike management systems, culture and leadership are softer aspects of workplace safety and health (WSH) management which may not be easy to evaluate and improve. Nevertheless, they are critical components of WSH management that have wide-ranging impact.

7.1 Introduction

Culture affects all management processes in an organisation because it is defined as the shared values and beliefs of the people in the organisation. A procedure or risk assessment (RA) may be very comprehensively written, but these detailed analyses or instructions can be easily defeated if the workers on the ground and the people in supervisory or management positions are not committed to them. In addition, leaders are the key people who can influence safety culture. This chapter will first discuss the definition of safety culture by introducing several safety culture models. Subsequently, safety leadership will be briefly introduced. Finally, the BP Texas City plant explosion case will be discussed to illustrate the importance of safety culture and leadership.

7.2 Defining Safety Culture

To define safety culture, we first need to understand organisational culture. One widely used definition of organisational culture is, "a system of shared values (*what is important*) and beliefs (*how things work*) that interact with a company's people, organisational structures, and control systems to produce behavioural norms (*the way we do things around here*)" (Uttal, 1983) (text in italics added by author). Culture can be very abstract and not clearly defined because it is made up of shared values and beliefs embedded in the minds of the people in the organisation. Culture can be observed through the pattern of behaviours across people, time and space.

Using the analogy of weather and climate, localised behaviours or specific events are like weather, which can vary in the short term. However, climatic patterns are relatively stable across longer spans of time and space. For instance, December is usually a very rainy month in Singapore, but it does not mean that there is rain every day in December. The longer-term pattern of

December being a wet month is similar to organisational culture, which is relatively stable across the years, though there are occasional deviations from the general pattern, meaning that individuals may still exhibit values, beliefs and behaviours not aligned with the organisation's culture from time to time.

Following the same analogy, although the weather in Singapore is relatively uniform across the small island of 721.5 km² (about 50 km by 27 km), Meteorological Service Singapore has, in the past, indicated that "rainfall is higher over the northern and western parts of Singapore and decreases towards the eastern part of the island". Likewise, organisational culture is not always uniform, even in small organisations. This is because each individual's values, beliefs and behaviours are always influenced by other factors such as the situation the person is in, the individual's experience and upbringing, personality, group influences, etc.

Applying the definition of organisational culture to safety, safety culture can be defined as a system of shared values and beliefs that interact with a company's people, organisational structures, and control systems to produce *safety-related* behavioural norms (see Figure 7.1). The behaviour of each individual person in the organisation is affected by the individual's characteristics, the safety culture (shared values and beliefs) of the organisation and the management system (structure and control system). Since all individuals are similarly influenced by the safety culture and management system, there will be similarities in the behaviours of all individuals in the organisation, but some deviation is expected due to differences in individual characteristics.

This definition of safety culture is aligned with the Event Causation Technique (ECT), which highlights that unsafe behaviour (a direct cause of incidents) is influenced by risk controls and management system components, safety culture and

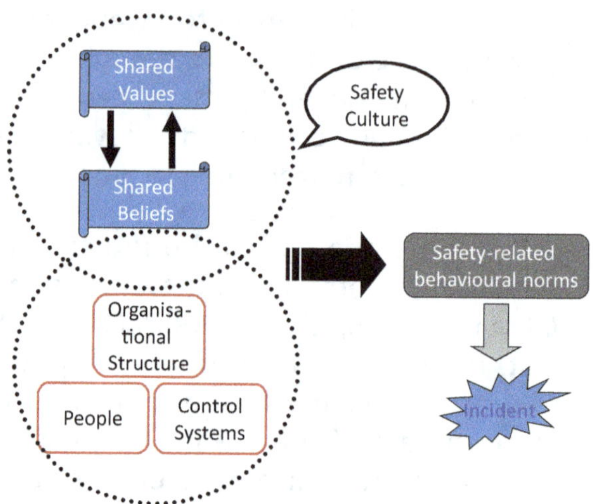

Figure 7.1 Safety culture

leadership. This concept is also highlighted by numerous researchers, for example Guldenmund (2007), who indicated that individuals' behaviours are influenced by the organisation's structure, culture and processes, which are "dynamically interrelated". He also indicated that an "organisation's culture cannot be isolated from its structure or processes".

To understand safety culture further, it is important to define beliefs and values. Beliefs refer to basic assumptions that a group of people hold in their minds. Some examples are, "safety is costly", "we have no time for safety and health issues", "accidents are a matter of luck" and "it all boils down to workers' behaviours; there is not much that the management can do". Beliefs drive behaviours. Hence, if one has beliefs that contradicts sound Workplace Safety and Health (WSH) management principles, their decision-making and intentions can be unsafe. Values refer to the relative importance of different concepts, e.g. WSH versus productivity and integrity versus ambition. Work-related values would be influenced by the beliefs and assumptions that we hold about the organisations that we work in/with and the people in the organisations. Shared values and shared beliefs are the common values and beliefs in a group of people. The common values and beliefs arise or were inculcated through shared experiences, stories and knowledge. The management system or norms in the group will help to create and maintain the culture in a workplace.

As culture is intangible, many managers do not focus on culture when they are managing the organisation, or they might not have the competency to influence culture. This can lead to management actions that fail to deliver or even backfire. It is important to design the management system to be aligned with the desired culture, while taking the current culture into consideration.

7.3 Different Layers of Culture

Based on previous research in organisational culture, Guldenmund (2000) presented a multi-layer model of safety culture (see Figure 7.2).

The core of safety culture is assumed to comprise shared basic assumptions, which are the shared beliefs presented earlier. These assumptions include the nature of reality and truth, time, space, human nature, human activity and human relationships. They are mainly implicit, but obvious or even factual to members of the organisation, yet they remain implicit, invisible and pre-conscious (i.e. assumed to be true in an unconscious fashion). These basic assumptions, or shared beliefs, have to be deduced from the middle and outer layers of culture.

The middle layer is made up of espoused *values or attitudes* regarding hardware, software, people and risks. Values or attitudes arise from basic assumptions or beliefs. This layer is more explicit and conscious, and it includes people's attitudes

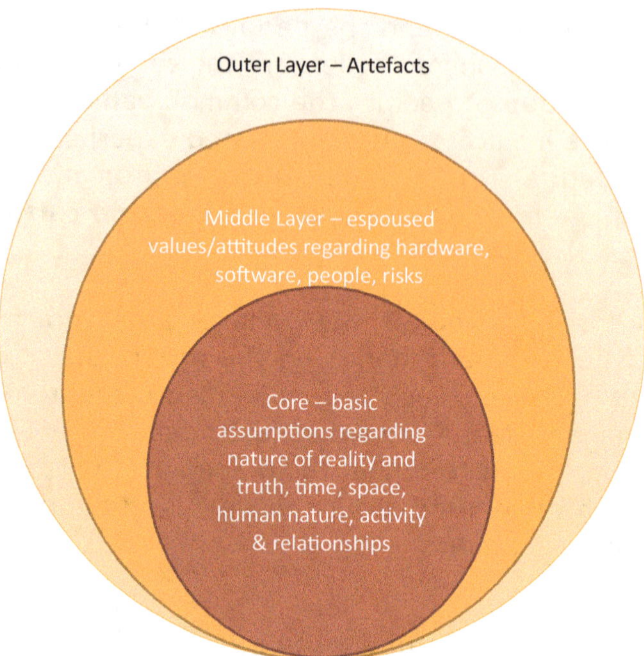

Figure 7.2　Levels of culture

towards objects or artefacts such as organisational policies, training manuals, procedures, formal statements, bulletins, accident and incident reports, job descriptions and minutes of meetings. Finally, the outer layer refers to the objects or artefacts that are visible and representative of safety culture, such as those mentioned earlier. Other examples of artefacts include management statements, meetings, inspection reports, dress codes, personal protective equipment (PPE), posters and bulletins. However, artefacts are poor representations of the underlying culture and are easily misinterpreted. For example, a company with many safety posters is not necessarily serious about the safety and health of its employees.

The model of different levels of safety culture indicated the difficulties in evaluating the safety culture of a workplace. Shared beliefs and values cannot be measured directly because many assumptions and beliefs are so deeply engrained that individuals may not be aware of them. Managers will have to rely on measurement of espoused values and attitudes and artefacts to judge the status of safety culture in their workplace. This will be discussed in more detail in this chapter and the next chapter.

7.4 Reason's Model of Safety Culture

Chapter 9 of Reason (1997) presents a model of safety culture which assumes that a positive safety culture is essentially an informed culture. An informed culture is, "one in which those who manage and operate the system have current knowledge about the human, technical, organisational and environmental factors that determine the safety of the system as a whole." A truly informed culture is in a sustained state of intelligent and respectful wariness. The organisation gathers the right kind of data using a safety information system. They do not just rely on lagging indicators; they also use leading indicators to help them manage safety, health, and environmental risks proactively. Reason (1997, p. 37) also highlighted, "... chronic unease is the price of safety. Studies of high-reliability organizations — systems having fewer than their 'fair share' of accidents — indicate that the people who operate and manage them tend to assume that each day will be a bad day and act accordingly." High-reliability organisations operate in highly complex environments and are expected to offer failure-free services. These organisations need to have robust WSH management systems supported by positive safety culture and a committed WSH leadership.

Reason (1997) felt that a safety culture can be socially engineered by implementing the different components of an informed culture. To him, an informed culture is based on four sub-cultures: reporting culture, just culture, flexible culture and learning culture. All four sub-cultures interact and come together to produce an organisation that has an informed culture.

7.4.1 Reporting culture

Reporting culture refers to the involvement and willingness of people to report hazards, errors and incidents. Possible disincentives to reporting culture include concerns about the impact of reporting, perceived additional work or trouble. To encourage reporting, there could be indemnity against disciplinary proceedings (as far as practicable) and ensuring confidentiality or de-identification of reporting staff.

There must be an independent and credible department or agency for disciplinary actions, which is separate from the department receiving reports of hazards and making WSH improvements. Other incentives include prompt feedback to the community about the actions and assessment of reported hazards or incidents and ensuring the ease of making the reports. The aviation industry has implemented successful reporting programmes by implementing many of these incentives for reporting. The success factor for a reporting system is trust in the system. Reporters must understand the importance of reporting and feel safe in reporting even their own errors.

7.4.2 Just culture

Just culture is essentially having a clearly established set of WSH rules, responsibilities, and accountabilities such that everyone in the organisation knows what is expected of them and accept the consequences that come with violating the WSH rules. Just culture is about the way that the organisation handles blame and punishment. A 'no-blame' culture is neither feasible nor desirable, as it may breed irresponsibility among some individuals, and this can spread across the organisation. Conversely, a punish-no-matter-what approach is also not feasible.

Any punishment needs to take the circumstances of the unsafe act into consideration. Even if the unsafe act was

Table 7.1   Effect of reward/punishment on behavioural change; adapted from Reason (1997)

	Immediate	**Delayed**
Reward	Positive effects	Doubtful effects
Punishment	Doubtful effects	Negative effects

intentional, the worker committing the unsafe act need not be culpable. Managers need to consider if the workers were encouraged to violate the rules due to lack of proper supervision or lack of effective training. It could also be a norm for everyone to break the rules because of the management's lack of interest in WSH. Decisions on the punishment also need to take into consideration if the same unsafe act would be committed if another reasonable man was placed in the same situation. In addition, the WSH track record of the worker needs to be factored in. These questions need to be considered prior to any punishment.

It is important to note that just culture promotes an atmosphere of trust, and helps to engender a reporting culture. There must be a good balance between punishment and rewards because rewards are known to produce positive effects in developing safe behaviour. With reference to Table 7.1, immediate recognition or other rewards are seen to be able to produce greater motivation to be safe. Some examples include giving workers small gifts or badges during site inspections as recognition of their positive safety behaviour. This is more effective in motivating safe behaviour than scolding and punishment.

7.4.3 Flexible culture

Flexible culture refers to the ability of an organisation to re-arrange its structure based on the needs of the work context.

Flexible culture allows people with the right expertise to provide input or take the lead when their expertise is needed. This is especially prominent in crisis situations where the manager or supervisors may not have the necessary expertise to manage the crisis. This may call for the fire safety manager or WSH manager to take over management of the site during the crisis. This can also happen during work, where those with different expertise are called upon to identify or control hazards. The ability to tap into expertise to suit the situation is a mark of flexible culture.

One of the key elements of flexible culture is the knowledge of where different expertise lie in the organisation. It is important for the safety information system to contain a comprehensive database of the strengths and competencies of different personnel within the organisation.

However, members of the organisation must also be willing to offer their expertise readily. Proactive involvement can be encouraged through incentives, but a sense of teamwork is more fundamental to a flexible culture.

7.4.4 Learning culture

Learning culture is the ability to learn from past incidents and available information to prevent future incidents. This involves a detailed analysis of information and having the ability to learn as a group. The ability to look beyond events and spot trends and patterns so as to prevent undesirable events is a key hallmark of a learning organisation. Learning culture requires an open mind and the willingness to listen to others.

However, an organisation need not have accidents to identify ways to improve their WSH management. The company can actively search for opportunities to learn by, for example, evaluating incidents from companies with similar operations, collecting feedback from workers periodically, reviewing audit reports, and analysing complaints from customers or other stakeholders. Another useful approach is to invite sister units or peer organisations to conduct benchmarking or peer review exercises to help the organisation review its performance and adopt best practices. It is also important to learn from successful departments or individuals within organisations and understand their key success factors. Improvement teams can be set up to innovate and focus on specific WSH issues. These improvement teams can be recognised and incentivised through competitions and awards.

7.5 Health and Safety Executive Safety Culture Questionnaire

Safety culture is made up of shared values and beliefs, which are generally intangible and they can be hard to measure and manage. Many researchers and organisations have developed questionnaires to help organisations assess safety culture. Many researchers use the term safety climate instead of safety culture because "safety climate" refers to the middle layer of safety culture (see Figure 7.2), i.e. attitudes, while safety culture refers to the core, which is "pre-conscious" and hard to measure. Many questionnaires use a quantitative approach, but some use a qualitative approach, combining responses from site personnel and the inspector or auditor so that safety culture improvement actions can be determined. One of these questionnaires is the Health and Safety Executive (HSE) safety culture questionnaire (Health and Safety Executive, n.d.), which is made up of seven key factors:

1. Management commitment
2. Communication
3. Employee involvement
4. Training/information

5.   Motivation
6.   Compliance with procedures
7.   Learning organisation

In contrast to the four sub-cultures of Reason's model, the HSE safety culture questionnaire highlights the importance of management commitment and the need for managers to be seen leading by example as far as safety is concerned. Another important aspect of management commitment is how managers allocate resources and how they make decisions when WSH and productivity or profit contradict each other. A good example of strong management commitment to safety is how Paul O'Neill, CEO of Alcoa (1987–2000) turned around the safety records of the company, and how he, while maintaining these impressive safety records, made Alcoa one of the most profitable companies in the world (Duhigg, 2014; Roth, 2012).

Several of the components of the HSE safety culture questionnaire are aligned with Reason's model. A learning organisation is synonymous with learning culture in Reason's model. Communication, employee involvement, information and motivation are related to reporting culture and just culture in Reason's model. However, the HSE safety culture questionnaire highlights the importance of training and compliance, which are not clearly highlighted in Reason's model. It is noted that some of the components of the HSE safety culture questionnaire, e.g. management commitment, communication, employee involvement, training/information, and compliance with procedures, are closely related to WSH management system elements described in the previous chapter. This is not surprising, because safety culture affects the way the WSH management system is established, implemented and maintained. A well-implemented WSH management system is reflective of a positive safety culture. The key difference is that the HSE safety culture questionnaire measures employees' perception of these

WSH management system elements as a proxy of the WSH beliefs and values of the organisation.

7.6 Zohar's Safety Climate Survey

Zohar (Hofmann *et al.*, 2017; Zohar, 1980; Zohar, 2010; Zohar and Luria, 2005) is known for his research on safety climate. Figure 7.3 shows an example of the questionnaire items in a typical safety climate questionnaire. Each statement will be rated by the respondent on a Likert scale of 1 to 5, or 1 to 7. The aggregated score will then give the management a sense of the safety climate of the organisation. In addition, it is possible to target specific management or supervisor behaviours to improve the safety climate. Zohar (2010) opined that safety climate is "a robust leading indicator or predictor of safety outcomes across industries and countries." The periodic measurement of an organisation's safety climate provides the organisation with another leading indicator, which must be read in conjunction with other leading and lagging indicators.

7.7 CultureSAFE and Others

The Singapore's Workplace Safety and Health Council (2015) has a comprehensive safety culture assessment and improvement programme called CultureSAFE. CultureSAFE provides the safety culture diagnostic tool, which is a questionnaire survey, and a range of recommended programmes to improve safety culture. Another well-known safety culture improvement programme is the Hearts and Minds programme (Energy Institute, n.d.), which was developed by Shell. The range of safety culture improvement programmes is very wide and each has a different take on what safety culture entails. Nevertheless, the different models agree that safety culture is about human and organisational aspects of WSH management. Safety culture influences

Organization-Level Safety Climate
Top management in this plant–company...

1. Reacts quickly to solve the problem when told about safety hazards.
2. Insists on thorough and regular safety audits and inspections.
3. Tries to continually improve safety levels in each department.
4. Provides all the equipment needed to do the job safely.
5. Is strict about working safely when work falls behind schedule.
6. Quickly corrects any safety hazard (even if it is costly).
7. Provides detailed safety reports to workers (e.g., injuries, near accidents).
8. Considers a person's safety behaviour when moving/promoting people.
9. Requires each manager to help improve safety in his/her department.
10. Invests a lot of time and money in safety training for workers.
11. Uses any available information to improve existing safety rules.
12. Listens carefully to workers' ideas about improving safety.
13. Considers safety when setting production speed and schedules.
14. Provides workers with a lot of information on safety issues.
15. Regularly holds safety-awareness events (e.g., presentations, ceremonies).
16. Gives safety personnel the power they need to do their job.

Note. Items cover three content themes: Active Practices (Monitoring, Enforcing), Proactive Practices (Promoting Learning, Development), and Declarative Practices (Declaring, Informing).

Figure 7.3   (*Continued*)

how the WSH management system and risk controls are established, implemented and maintained.

7.8 Safety Leadership

This section will briefly discuss transformational leadership and transactional leadership in the context of WSH. One of the

Group-Level Safety Climate

My direct supervisor…

1. Makes sure we receive all the equipment needed to do the job safely.
2. Frequently checks to see if we are all obeying the safety rules.
3. Discusses how to improve safety with us.
4. Uses explanations (not just compliance) to get us to act safely.
5. Emphasizes safety procedures when we are working under pressure.
6. Frequently tells us about the hazards in our work.
7. Refuses to ignore safety rules when work falls behind schedule.
8. Is strict about working safely when we are tired or stressed.
9. Reminds workers who need reminders to work safely.
10. Makes sure we follow *all* the safety rules (not just the most important ones).
11. Insists that we obey safety rules when fixing equipment or machines.
12. Says a "good word" to workers who pay special attention to safety.
13. Is strict about safety at the end of the shift, when we want to go home.
14. Spends time helping us learn to see problems *before* they arise.
15. Frequently talks about safety issues throughout the work week.
16. Insists we wear our protective equipment even if it is uncomfortable.

Note. Items cover three content themes: Active Practices (Monitoring, Controlling), Proactive Practices (Instructing, Guiding), and Declarative Practices (Declaring, Informing).

Figure 7.3   Questionnaire items developed by Zohar and Luria (2005)

key leadership approaches adopted by many leaders to influence subordinates' safety behaviour is the use of reward and punishment systems. This is known as transactional leadership (Northouse, 2016). In contrast, transformational leadership "evokes changes in subordinates' value systems to align them with organizational goals" (Clarke, 2013) and numerous studies have shown that it has positive effects on the safety behaviour and safety participation of subordinates. This section will highlight how leaders can adopt the transformational leadership style so as to engender positive safety culture in their workplace. Another popular leadership model relevant to our discussion is Situational Leadership® II (SLII®) (Blanchard, 1985; Blanchard *et al.* 2013; Northouse, 2016), which will be discussed in this section.

7.8.1 Transformational leadership

Many leadership theories were developed over the years, but since Downton (1973) coined the term *Transformational Leadership*, it has become one of the most widely researched leadership theories (Northouse, 2016). So what is *Transformational Leadership*? It is widely accepted that transformational leadership encompasses the following distinct but correlated dimensions (aka the 4 '*I*'s):

1. Idealised influence, i.e. leader instils confidence and behaves in admirable ways that cause the followers to identify with him/her
2. Inspirational motivation, i.e. leader inspires others towards goals, provides meaning, optimism and enthusiasm, and articulates a vision that is appealing and inspiring to others
3. Intellectual stimulation, i.e. leader challenges assumptions, takes risks and encourages subordinates to be creative
4. Individualised consideration, i.e. leader shows interest in subordinates' personal and professional development and listens to followers' needs and concerns

At the same time, much research has shown that WSH leadership is one of the key criteria of a positive safety climate or culture (Zohar, 2010). In essence, "leaders create climate" (Lewin *et al.*, 1939). To facilitate ease of discussion, climate is deemed to be a proxy to culture in this section, i.e. they are treated as "pretty much the same thing". In the case of transformational leadership, it is quite intuitive that transformative leadership will have a positive impact on safety culture. A closer look at each of the 'I's in transformational leadership will shed more light on the relationship between transformational leadership and safety culture (adapted from Clarke (2013)):

1. Idealised Influence: managers adopting transformational leadership are role models by doing what is morally right as compared to focusing on production goals only. Idealised influence encourages focus on WSH and sustainable ways of working, in contrast to prioritising short-term benefits due to work pressure. Leaders high in idealised influence will demonstrate their personal commitment to WSH as a core value. The personal commitment will improve followers' trust in management and loyalty, which can lead to improvement in overall performance.

2. Inspirational motivation: transformational leaders motivate followers to be part of the organisation's shared vision and to go beyond their individual needs. In terms of WSH, transformational leaders inspire their followers to achieve safety levels previously believed to be impossible, e.g. Vision Zero or Zero Harm. Transformational leaders frequently use symbols and stories to articulate their vision and inspire followers.

3. Intellectual stimulation: Intellectual stimulation arises when leaders encourage followers to address WSH issues and enhance information sharing by challenging long-held assumptions and promoting innovation in WSH. This engenders a learning culture, where followers become more aware of and better understand WSH problems, encouraging innovative, rather than reckless solutions.

4. Individualised consideration: transformational leaders have a strong and sincere interest in their followers' well-being, which naturally includes WSH. Transformational leaders are not satisfied with compliance with WSH legislations because they want to ensure that their followers do not experience injuries and ill-health.

The positive impact of transformational leadership on safety culture and safety performance is supported by empirical studies. For example, Barling *et al.* (2002) studied 174 restaurant workers and 164 young workers from diverse jobs using selected questions from the Multifactor Leadership Questionnaire (MLQ) (Bass and Avolio, 1990), a shortened version of the safety climate questionnaire by Zohar (1980), and they also measured the self-reported safety consciousness and safety incidents. The study showed that, among other factors, safety-specific transformational leadership plays a significant role in influencing "safety climate, safety consciousness, and safety-related events." On the other hand, passive leadership is shown to be related to poorer safety climates (Kelloway *et al.*, 2006). Thus, past studies have shown that leadership influences climate, and in turn, climate had been shown to influence the occurrence of workplace injuries (Huang *et al.*, 2012). Transformational leadership is a potentially powerful tool to influence safety culture and WSH.

Based on the 4 'I's, leaders can adopt the following practices to be more aligned with transformational leadership (Kouzes and Posner, 2002):

1. Model the way: Be clear about one's own values and philosophy and be a role model.
2. Inspire a shared vision: Develop a compelling vision that helps followers visualise positive outcomes and see how their dreams can materialise.
3. Challenge the process: Transformational leaders are pioneers because they are willing to experiment, try new things and learn from mistakes so as to achieve better results.
4. Enable others to act: Outstanding leaders are effective at working with people by building trust, promoting collaboration and listening respectfully. They empower others to make decisions and feel good about their work by linking it with greater purposes.

5. Encourage the heart: Be attentive to followers' need for recognition and reward. Show appreciation and encouragement to others.

Transformational leadership has the potential to improve the performance of organisations, including safety culture and hence WSH performance. Positive changes in followers' behaviour will take time, but the persistent implementation of transformational leadership will reap results in WSH and organisational performance in general. This is not saying that it is the only leadership style that is applicable to all situations. Successful leaders will have the wisdom to observe and apply suitable leadership actions in different situations so as to improve WSH. In the end, leaders have to take it upon themselves to use transformational leadership or other approaches to convince their followers to adopt safer work methods.

7.8.2 Situational Leadership®

The basic premise of the SLII® model (Blanchard, 1985; Blanchard et al., 2013; Northouse, 2016) is that leaders need to adapt his/her leadership style and behaviour based on the development level of his/her followers. The leadership dimension is simplified into a dichotomy of directive and supportive. The followers' development levels are categorised by their competency and commitment to assigned goals. Followers' competency and commitment vary across time and situations, so the leader will have to match the leadership approach to suit the followers' collective level of competency and commitment.

A directive leadership behaviour involves one-way communication of the goals, methods to conduct the task, ways to evaluate the performance etc. fron the leader to the follower. On the other hand, a supportive leadership behaviour uses two-way communication to show social and emotional support to followers.

Table 7.2   Leadership approach based on SLII® model (Blanchard, 1985; Blanchard *et al.* 2013; Northouse, 2016)

Leadership Approach	Leadership Style		Followers' Development Level	
	Directive	Supportive	Competence	Commitment
Directing	High	Low	Low	High
Coaching	High	High	Low to some	Low
Supporting	Low	High	Moderate to high	Variable
Delegating	Low	Low	High	High

Within the context of WSH management and based on Table 7.2, leaders should use the suitable safety leadership approach when working with different types of followers. If the followers do not have suitable competency in WSH management, but they are highly committed to improving WSH, the leader should focus on communicating WSH goals, giving detailed instructions on the methods to be used by the followers and then supervising them carefully (Northouse, 2016). Coaching is suitable when the follower has low or some competence in and is low in commitment to WSH. In this situation, besides being directing, the leader needs to be supportive of the followers through consultation, solving problems together, and encouraging the followers. Supporting is used when the followers have moderate to high competence in WSH management, but have variable commitment. In this situation, the leader focuses on trying to motivate the followers to contribute their best skills. The leader will have to listen, feedback and praise the followers appropriately. The followers will be directly in-charge of day-to-day implementation of WSH management tasks, but the leader will still be involved in goal setting and problem solving. Last, when the followers are high in competence and commitment, the leaders aim to get the group to agree on the goals and directions and let followers be responsible for achieving the goals based on what they think is suitable.

The SLII® provides useful WSH leadership guidance to frontline leaders and top management alike. A supervisor can assess each of his workers based on competence and commitment to decide how to direct or support the worker. Similarly, from the perspective of an organisation, the top management will have to assess the level of WSH competency and commitment of its employees in general, which is an indication of the safety culture of the organisation, and then decide what leadership approach is needed to improve safety culture as a whole or use different approaches for different groups within the organisation.

7.9 BP Texas Explosion Case Study

The BP Texas City Refinery explosion and fire (Chemical Safety Board, 2007) is one of the worst industrial accidents in recent history. The accident happened on 23 March 2005 at 1:20pm (GMT-5). The explosion and fire killed 15 people and injured another 180. The financial losses were estimated to be at least US$1.5 billion. The explosion occurred during the start-up of an isomerisation (ISOM) unit, when a raffinate splitter tower was overfilled. The overfilling led to the opening of pressure relief devices, but it resulted in a flammable liquid geyser (a fountain of high pressure jets of fluid and gas that shoots into the air) from a blowdown stack that did not have a flare (a safety device which can burn off gas or liquid released through the blowdown). This resulted in the formation of a flammable vapour cloud that was ignited by the idling engine of a pick-up truck parked near the blowdown stack.

The investigations conducted after the accident revealed severe organisational and safety culture problems in BP Texas City Refinery. Some of them are provided below:

1. The BP Board of Directors did not provide effective oversight of BP's safety culture and major accident prevention

programmes. The Board did not have a member responsible for assessing and verifying the performance of BP's major accident hazard prevention programmes. *Note: This shows that safety leadership was inadequate. An informed culture where senior management were in the know of safety issues was not present.*

2. Numerous surveys, studies, and audits identified deep-seated safety problems at Texas City, but the response of BP managers at all levels was typically "too little, too late." *Note: This is a sign of lack of proactive actions by managers and leaders, i.e. a safety leadership issue.*

3. Cost-cutting, failure to invest, and production pressures from BP Group executive managers impaired process safety performance at Texas City. *Note: This reflected how cost-cutting was valued over process safety; a sign of poor safety culture and a lack of safety leadership.*

4. A "check the box" mentality was prevalent at Texas City, where personnel completed paperwork and checked off on safety policy and procedural requirements even when those requirements had not been met. *Note: Routine violation was common and such behaviour is again a reflection of unsafe shared values and beliefs. "Check the box" is the same phenomenon as "paper exercise" highlighted in Chapter 4. These phenomena are reflections of surface compliance (Hu et al., 2020), which is compliance in form, but not in intent. In contrast, deep compliance is focused on compliance of the intent (Hu et al., 2020).*

5. BP Texas City lacked a reporting and learning culture. Personnel were not encouraged to report safety problems and some feared retaliation for doing so. The lessons from incidents and near-misses, therefore, were generally not captured nor acted upon. Important relevant safety lessons from a British government investigation of incidents at BP's Grangemouth, Scotland refinery were also not incorporated at Texas City. *Note: This is directly aligned with*

Reason's (1997) model of safety culture, where reporting and learning culture are critical components.

6. Reliance on the low personal injury rate at Texas City as a safety indicator failed to provide a true picture of the processes' safety performance and the health of the safety culture. *Note: This is a poor choice of WSH indicator, which reflects an inadequate WSH management system. There is a lack of suitable indicators to reflect the process safety of the plant.*

7. Safety campaigns, goals, and rewards focused on improving personal safety metrics and worker behaviours rather than on process safety and WSH management systems. While compliance with many safety policies and procedures was deficient at all levels of the refinery, Texas City managers did not lead by example regarding safety. *Note: Similar to the above point, this is a major management system issue.*

The above findings showed that an inadequate WSH management system, poor safety culture and inadequate safety leadership were intertwined and they contributed to the BP Texas explosion. These underlying factors are almost always found in major accidents, albeit with variances in the details. For example, the investigation into the flooding of the Bishan train tunnel, which led to a 20-hour train service disruption in Singapore on Oct 7, 2017 (Kuek, 2017), showed that even though the flooding was triggered by heavy rain, the incident was a result of a poorly implemented maintenance system (including falsified maintenance records) contributed by "deep-seated cultural issues". Then Transport Minister Khaw Boon Wan said that, "a positive corporate culture starts from the top" (Tan, 2017). At the same time, he emphasised that, "growing the right culture is the responsibility of everyone — from the top leadership down to the workers." This is definitely true because culture is the product of shared values and beliefs, and everyone has an influence over culture, including safety culture, but top management are expected to be directly accountable for culture.

7.10 Learning Disabilities and Safety Culture

Reason (1997) highlighted the importance of learning culture as a sub-culture of an informed culture and safety culture. At the same time, WSH management standards such as ISO 45001:2018 (see Chapter 6) are based on the Plan-Do-Check-Act (PDCA) cycle, which emphasises the importance of continual improvement and learning. Senge (2006, p. 4) highlighted that, "[t]he organizations that will truly excel in the future will be the organizations that discover how to tap people's commitment and capacity to learn at all levels in an organization". An organisation that "truly excels" must have a strong safety culture to sustain WSH performance across time. However, many organisations have learning disabilities that can lead to poor learning and safety culture. Senge (2006) highlighted seven learning disabilities:

1.   I am my position
Over time, individuals become attached to their positions, identifications and roles, and this leads to a "silo mentality", where they are no longer concerned with or become unable to see how their actions affect the other parts of their organisation and vice versa. When a procurement executive is asked to consider the safety and health of workers, he cannot see why he needs to do so because to him it is the workplace safety and health officer's (WSHO) job to do so, and not his. The procurement executive did not realise that many workplace hazards are introduced by the procurement department when unsafe equipment or incompatible material are purchased. A myopic view of roles is common, and people do this to protect themselves from overwork and to keep roles and responsibilities clear and systematic. There is always a concern that once they help "others" to do their work, their job scope will become enlarged and unmanageable. Such a

mentality focuses on what people do and not the purpose of the organisation that they are in. This learning disability limits the possible actions that organisations can take when facing complexity and the full potential of an organisation can never be fully utilised.

2.   The enemy is out there
I had observed parents who comfort their crying child who had just hit the wall by slapping the wall and saying, "ok, I'm punishing the wall for hitting you on the head. Don't cry anymore..." Many people have the tendency to blame others for the undesired situation they are in, but a friend of mine was extraordinary. We were trying to fix something on the ceiling and the ladder we had was too short, so the author said, "the ladder is too short." The friend replied in a philosophical tone, "No, I am too short." Such a mentality is remarkable. Both of us were not wrong, but his response showed a fundamental shift in mindset. He looks for his own contribution to the problem we were facing, and such a mentality allows organisations to look inward for ways to deal with complex problems and to achieve the desired outcomes.

An example provided by Wright (Meadows and Wright, 2008) further illustrates the learning disability of "the enemy is out there". She first shows her class how a Slinky bounces up and down when it hangs from her hand. She asked her students, "What made the Slinky bounce up and down like that?" The students said it was her hand that caused the Slinky to bounce. When Wright hangs a small box from her hand, the box does not bounce like the Slinky. Both the Slinky and the box were placed in the same situation (hung from her hand), but the responses were drastically different. The internal structures of the Slinky and the box led to the differences in behaviour. Internal structure refers to the way the different elements or components of a system are connected, e.g. the way materials are joined to form the Slinky and the box.

Similarly, a manager can complain that the client is the one who is causing all the WSH problems by not putting enough emphasis on WSH. This could lead to the manager's failure to identify ways that he and his team can better manage their WSH management system and culture.

3. The illusion of taking charge

Being proactive in doing "something" might actually be reactiveness in disguise. When a manager starts to be proactive and take charge when dealing with safety culture problems, he might take an aggressive stand against an "enemy out there". This would worsen the problem of "the enemy is out there". For example, the manager might think that it is the subcontractors' fault for not complying with WSH rules given to them, and he did not consider the need to review their contractor selection process and other related processes to assess how they could have selected better contractors to assist in their projects.

4. The fixation on events

As explained in earlier chapters, events are episodes of significance. When managers are fixated on events, it means that they do not look for the underlying structure or factors that influenced the occurrence of an event. When a project has an accident, it is easy to blame it on the worker who caused the accident, but, as highlighted in Chapter 3, it is more important to understand the management system's inadequacies, safety culture and safety leadership issues that led to the accident. Focusing on events leads to event level explanations and solutions, such as firing the WSHO or blaming individual workers, which do not help the organisation to learn and improve.

5. The parable of the boiled frog

The parable of the boiled frog is about our insensitivity to gradual changes in our environment. The parable says that if a frog is placed in a pot of water and the temperature of the water is raised slowly, the frog will do nothing to escape the impending danger. It might start to feel uncomfortable at some point, but by the time the frog notices the danger of the rising temperature, it is too late, and it gets boiled alive. This sounds very much like the climate change challenges that the world is experiencing. From a WSH point of view, it is important for managers to set up a monitoring system to continuously assess the changes in the organisation and its environment. One of the common mistakes managers make is to belittle possible WSH issues raised by team members or stakeholders only to realise later on that those issues lead to WSH incidents (as will be observed in Chapter 10). Safety culture changes progressively across time and it is important for managers to monitor safety culture and be proactive in maintaining or improving safety culture. Consistent and deliberate considerations can help to prevent the organisation from becoming a frog in boiling water.

6.   The delusion of learning from experience

We learn best from experience, but that is only if we can see the consequences of our actions. If there is significant delay between action and consequence, or if the consequences of our actions are not known to us, then it is difficult to learn from such experience. For example, a manager might not have been communicating with his team frequently, and he will not know that his lack of communication is slowly causing him to be alienated from his team. This further deteriorates communication between the manager and his team, leading to a vicious cycle where lack of communication leads to even less communication in the team. The manager might think that by cutting down on meetings he had successfully improved the productivity of his team, but in fact, the lack of meeting has unintended effects on his team. However, the manager will only experience the negative consequences after some time, which becomes a barrier to him realising his mistakes.

7.   The myth of the management team

The management team is supposed to be a team of individuals who work together to bring the project to success. However, the reality is that most teams in workplaces are not performing as a team. Instead, individuals might be more concerned with individual key performance indicators (KPIs), and office politics and turfs. Many teams do not communicate well and break down under pressure arising from complexity. Dysfunctional teams are not able to decipher complexity and derive solutions to handle the complexity of the work, and that can result in WSH incidents. Successful management teams have individuals who are willing to let go of their egos and territories and explore complex problems humbly with their team members. They are open minded and willing to listen to others. Such teams are hard to come by, but extremely valuable and powerful when handling complex projects.

These seven learning disabilities lead to organisations being unable to manage hazards and risks arising from work, leading to injuries and ill health. On the other hand, organisations "where people continually expand their capacity to create the results they truly desire, where new and expansive patterns of thinking are nurtured, where collective aspiration is set free, and where people are continually learning how to learn together" (Senge, 2006, pp.3–4) are learning organisations that can excel and achieve excellent WSH performance. Senge (2006) further postulated that there are five fundamental disciplines critical to the development of a learning organisation. This ensemble of five disciplines will be discussed in the next chapter.

7.11 Conclusions

Safety culture is about an organisation's shared values and beliefs that influence safety related behaviours. There are various models of safety culture and some were discussed in this chapter. Regardless of the definitions and models, the key thing to note is that safety culture influences, albeit not in a deterministic manner, human behaviours while implementing

WSH procedures and policies. If there is no buy-in from the people implementing the WSH management system, the system will never work. When dealing with people, we need to use humanistic language and be in sync with emotions. Despite the importance of legislations, liabilities and statistics, if there is an over-focus on them, it is not possible to get WSH messages across. Leaders and managers have to demonstrate genuine care for the workers to let the workforce know that everybody needs to use and improve the WSH management system. Thus, safety leadership is intertwined with safety culture, and the key to a positive culture rests with top management.

A learning organisation, as proposed by Senge (2006), is aligned with a positive safety culture. The seven learning disabilities highlighted by Senge (2006) are some of the possible barriers to a achieving a learning culture, which is essential for a positive safety culture.

Review Questions

1. Explain Reason's model of safety culture and contrast it with the HSE Safety Culture questionnaire.
2. Why is top management so important in improving safety culture?
3. What can top managers do to demonstrate their commitment to WSH?
4. Explain the difference between safety culture and safety climate.
5. Differentiate transactional and transformational leadership.
6. Describe the different leadership approaches in the Situational Leadership® II model.
7. List some of the safety culture problems at BP Texas City Refinery prior to the accident in 2005.
8. Explain how the seven learning disabilities described in Senge (2006) contradicts the formation of the positive safety culture described in Reason (1997).

References

Barling, J., Loughlin, C., and Kelloway, E. K. (2002). "Development and test of a model linking safety-specific transformational leadership and occupational safety." *Journal of Applied Psychology*, 87(3), 488.

Bass, B. M., and Avolio, B. J. (1990). *Transformational leadership development: Manual for the multifactor leadership questionnaire*, Consulting Psychologists Press.

Blanchard, K. H. (1985). *SLII®: A situational approach to managing people*. Escondido, CA: Blanchard Training and Development.

Blanchard, K., Zigarmi, P., and Zigarmi, D. (2013). *Leadership and the one minute manager: Increasing effectiveness through Situational Leadership® II*. New York: William Morrow.

Chemical Safety Board (2007). "Investigation Report — Refinery Explosion and Fire (BP Texas)." http://www.csb.gov/. (accessed December, 2008).

Clarke, S. (2013). "Safety leadership: A meta-analytic review of transformational and transactional leadership styles as antecedents of safety behaviours." *Journal of Occupational and Organizational Psychology*, 86(1), 22–49.

Downton, J. V. (1973). *Rebel leadership: Commitment and charisma in the revolutionary process*, New York: Free Press.

Duhigg, C. (2014). *The power of habit: Why we do what we do in life and business*. New York: Random House.

Energy Institute (n.d.). "Hearts and Minds." http://www.eimicrosites.org/heartsandminds/. (accessed Oct 8, 2015).

Guldenmund, F. W. (2000). "The nature of safety culture: A review of theory and research." *Safety Science*, 34(1–3), 215–257.

Guldenmund, F. W. (2007). "The use of questionnaires in safety culture research — an evaluation." *Safety Science*, 45(6), 723–743.

Health and Safety Executive (n.d.). "Common topic 4: Safety culture." www. hse.gov.uk/humanfactors/topics/common4.pdf. (accessed September 2, 2020).

Hofmann, D. A., Burke, M. J., and Zohar, D. (2017). "100 Years of occupational safety research: From basic protections and work analysis to a multilevel view of workplace safety and risk." *Journal of Applied Psychology*, 102(3), 375–388.

Hu, X., Yeo, G., & Griffin, M. (2020). "More to safety compliance than meets the eye: Differentiating deep compliance from surface compliance." *Safety Science*, 130, 104852. doi:10.1016/j.ssci.2020.104852

Huang, Y.-H., Verma, S. K., Chang, W.-R., Courtney, T. K., Lombardi, D. A., Brennan, M. J., and Perry, M. J. (2012). "Supervisor vs. employee safety perceptions and association with future injury in US limited-service restaurant workers." *Accident Analysis & Prevention*, 47, 45–51.

Kelloway, E. K., Mullen, J., and Francis, L. (2006). "Divergent effects of transformational and passive leadership on employee safety." *Journal of occupational health psychology*, 11(1), 76.

Kouzes, J. M., and Posner, B. Z. (2002). *The leadership challenge*. San Francisco: Jossey-Bass.

Kuek, D. (2017). *In full: SMRT CEO Desmond Kuek on 'deep-seated cultural issues' behind history of service disruptions*, https://www.todayonline.com/singapore/smrt-ceo-desmond-kueks-statement-response-oct-7-tunnel-flooding-nsl. (accessed Apr 6, 2020)

Lewin, K., Lippitt, R., and White, R. K. (1939). "Patterns of aggressive behavior in experimentally created 'social climates'." *The Journal of Social Psychology*, 10(2), 269–299.

Meadows, D. H., and Wright, D. (2008). *Thinking in systems: A primer*. White River Junction, Vt: Chelsea Green Pub.

Northouse, P. G. (2016). *Leadership: Theory and practice*, SAGE, Thousand Oaks.

Reason, J. (1997). *Managing the risks of organizational accidents*. Aldershot: Ashgate.

Roth, M. (2012). "'Habitual excellence': The workplace according to Paul O'Neill." http://www.post-gazette.com/business/business-news/2012/05/13/ Habitual-excellence-The-workplace-according-to-Paul-O-Neill/stories/201205130249. (accessed Oct 8, 2015).

Senge, P. (2006). *The fifth discipline — The art & practice of the learning organisation*. New South Wales: Random House Australia.

Tan, C. (2017). "Parliament: Right culture starts from top, says Khaw Boon Wan." https://www.straitstimes.com/singapore/transport/parliament-right-culture-starts-from-top-says-khaw. (accessed Apr 6, 2020).

Uttal, B. (1983). "The corporate culture vultures." *Fortune*, Oct. 17, 66–72. Workplace Safety and Health Council (2015). "Apply for CultureSAFE." http://www.wshc.gov.sg/. (accessed Oct 8, 2015).

Zohar, D. (1980). "Safety climate in industrial organizations: Theoretical and applied implications." *Journal of Applied Psychology*, 65(1), 96–102.

Zohar, D. (2010). "Thirty years of safety climate research: Reflections and future directions." *Accid. Anal. Prev.*, 42(5), 1517–1522.

Zohar, D., and Luria, G. (2005). "A multilevel model of safety climate: Cross-level relationships between organization and group-level climates." *Journal of Applied Psychology*, 90(4), 616–628.

CHAPTER 8

Improving Safety Culture

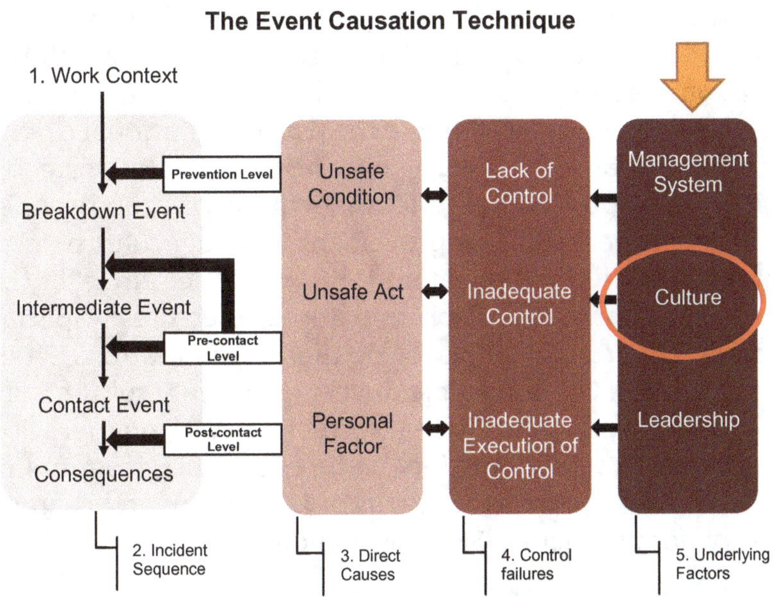

As established in the earlier chapters, safety culture interacts with management systems to influence behavioural norms and hence incident occurrence. To improve safety culture, the organisation needs to undergo fundamental changes, which can be implemented based on organisational change theories and concepts.

8.1 Introduction

Organisational change is defined as a change to the *core* of an organisation. The *core* of an organisation includes vision, mission, strategies, culture, structure, competencies, critical management system processes, and personnel. This chapter provides a brief overview of how organisational change management concepts can be used to improve safety culture. Managers, project managers, engineers, workplace safety and health (WSH) managers, and WSH officers will need to lead safety culture improvement initiatives, which constitutes a significant organisational change.

In the context of WSH, change management usually refers to management of change (MOC) (see Chapter 6), which is an operational level change that is systematically managed using a risk-based approach to minimise the risk of operational changes leading to hazards that cause accidents or ill health. Instead of MOC, this chapter will be discussing change at the organisational level, specifically, improvement in safety culture. In the context of WSH, the motivation for improvement in safety culture is frequently a reaction to external pressures, especially enforcement from a regulator or client organisations after a major accident. Other possible drivers for a safety culture initiative include new leadership that focuses on WSH, merger with a company with stronger safety culture, pressure from unions, new WSH management standards or legislation, and changing industry norms where WSH is given more emphasis.

After a major accident, an organisation's survival and future success depend on its ability to improve its safety culture and WSH management system in response to internal and external pressures. If an organisation is not able to demonstrate its ability to ensure workers' safety and health, they may be put out of business and managers and leaders may be removed from their positions. Thus, organisations need to proactively improve their safety culture and WSH management system, and not wait for a major incident to provide the impetus to do so. However, most organisational change initiatives fail, so it is important for us to understand how to design and implement successful organisational change. In this chapter, the key focus will be on approaches to improve safety culture, which is fundamental to an effective WSH management system.

8.2 Improving Safety Culture

Using the definition of safety culture introduced in the last chapter, three possible approaches to improve safety culture can be identified (see Figure 8.1).

First, a direct intervention approach focuses on the middle layer of safety culture, i.e. espoused values and attitudes. In this approach, management aims to express, communicate and diffuse desired safety beliefs and values throughout the organisation so that all employees share the same vision of a safe and healthy workplace. This can be done through facilitated focus group discussions, interviews and safety climate surveys. Focus group discussions and interviews will produce qualitative cases or stories on what employees see as anecdotal evidence of

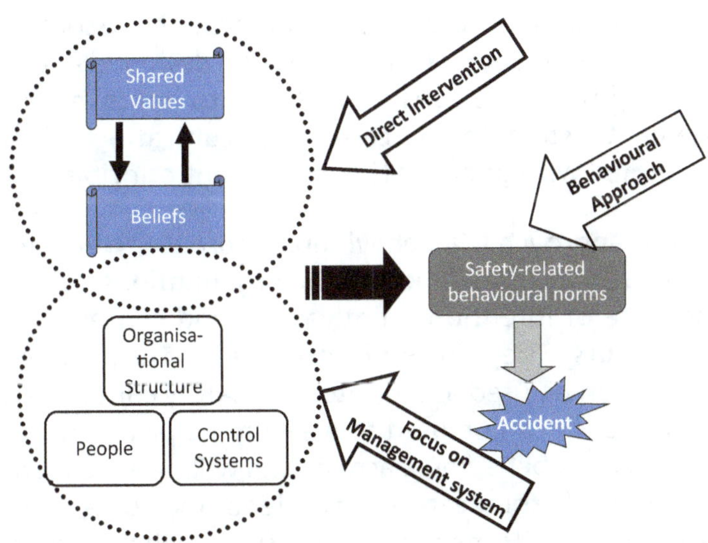

Figure 8.1   Approaches to improve safety culture

shared values and beliefs. These stories can be very powerful in further establishing shared values and beliefs. The effectiveness of directly discussing safety culture is highly dependent on the quality of communication and facilitation. A safety climate survey provides a quantitative source of information indicative of the shared values and beliefs that can be used to triangulate along with the information obtained from focus group discussions and interviews. The direct intervention approach assumes that safety culture can be verbalised, discussed and modified through discussions.

Second, safety culture can be improved by focusing on the WSH management system. This approach is an outside-in approach. Since management standards (e.g. ISO 45001 and ISO 14001) are essentially good practices adopted from organisations with positive safety culture, it is assumed that organisations that model themselves after these good practices will develop a good safety culture. This is focused on the outer layer of safety culture, but since a management system includes training, communication and consultation, there will be overlaps with the direct intervention approach. The assumption is that when the management system is well-designed and implemented, and there are consistent checks to improve the WSH management system, the process will be cultivating values and beliefs aligned with good WSH management principles.

The last approach is a behavioural approach, which focuses on the behaviours of members of the organisations. Behaviours arise because of intentions, motivations and consequences of the behaviours. They are more observable than values and beliefs, and can be used as indicators for safety culture. A quantitative approach can be used to measure behaviours. Tracking and trending the behaviours across time will provide managers with data which they can use as a leading indicator of the effectiveness of WSH management. During behavioural observations, the observer will be trained to provide positive reinforcement or feedback to motivate the person being observed to work safely. Such an approach is known as the behaviour-based safety (BBS) approach and there are many variations to how BBS is implemented. BBS assumes that external motivation will change behaviour in the long run, and as a result safety culture is improved.

All three approaches are overlapping, viable and inter-dependent. An effective safety culture improvement initiative will involve the use of all three approaches in different stages of the programme.

8.3 Cultivating Safety Culture through Disciplines of Learning Organisation

As highlighted in Chapter 7, Senge (2006) emphasised the importance of a learning organisation in coping with changes and complexities. Safety culture improvement initiatives, like other organisational changes, are always complex because of the uncertainty, ambiguity, and complex interdependence between people and tasks. According to Senge (2006), the discipline of systems thinking helps an organisation to understand complexity, but for an organisation to become a learning organisation, it needs to practise four other disciplines that can be grouped into two groups. The first group is "aspirations", which includes personal mastery and shared vision. The second group is "reflective conversation", which includes mental models and team learning and dialogue. All four disciplines are related to the direct intervention approach, in particular, shared vision, and team learning and dialogue. The following gives an overview of each of the disciplines.

Senge (2006) considers systems thinking, which had been discussed across the different chapters, as the cornerstone of a learning organisation. Systems thinking is a way of seeing systems (including WSH management systems) holistically. A systems thinker would strive to understand the underlying patterns, structures and factors influencing the occurrence of WSH incidents, so that more fundamental solutions can be derived to achieve sustained positive WSH performance even in complex and challenging environments. We have introduced systems thinking tools such as causal loop diagrams and system archetypes; these are useful tools to evaluate and improve safety culture, but the most important contribution of systems thinking to safety culture is the proactive mindset of wanting to understand how our actions contribute to the systemic problems we face. In seeking to understand the system, we need to look at how different elements of the system, including ourselves, interact to produce the behaviours we observe. This holistic and inward-looking approach changes the fundamental beliefs and values of individuals, i.e. organisational and safety culture will be changed.

Personal mastery (Senge, 2006, p. 7) "is the discipline of continually clarifying and deepening our personal vision, of focusing our energies, of developing patience, and of seeing reality objectively". An organisation is made up of individuals and a learning organisation can only be formed if the individuals in the organisation are truly committed to learning. In our context, individuals must learn how to make the workplace safe and healthy and this commitment must be encouraged and facilitated by the organisation. Personal mastery is essentially about employees learning to create and sustain the "creative tension" or resist the pressure to give up on their vision when facing the current reality. This is related to the archetype of "eroding or shifting goals", where individuals and their organisation must "hold the vision" even when the current reality is way below the vision.

Mental models (Senge, 2006, p. 8) "are deeply ingrained assumptions, generalizations, or even pictures or images that influence how we understand the world and how we take action." Based on our definition of safety culture, shared mental models of WSH is essentially the safety culture of an organisation. In the context of WSH, the discipline of mental models is about looking inwards to understand our basic assumptions about WSH (e.g. our assumptions about how WSH incidents happen) and evaluate them. This can then be the basis for "learningful" conversations that balance inquiry and advocacy.

The discipline of building shared WSH vision is about safety leadership. Good WSH leaders inspire and sustain a vision of an organisation that is injury and illness free. They rally the people to be committed to WSH goals, values and mission. The shared vision must be truly shared by the individuals in the organisation and, as discussed in Chapter 7, different leadership approaches can be adopted to build the shared vision. However, it must be noted that leadership does not only include the top management. Anyone in the organisation can be a WSH leader by being committed to the WSH-related activities and lead through their actions. A shared vision is critical to a safety culture improvement initiative because it helps the organisation to focus their energy and resources and it establishes a reference point for everyone in the organisation. Safety culture, being defined as shared values and beliefs about WSH, is directly related to a shared vision. If the employees of an organisation are truly committed to the same vision of zero injury and illness, they will naturally value WSH and will believe that they need to do more to achieve the shared vision. The process of building a shared vision involves individuals having their personal vision through the discipline of personal mastery.

Building a shared vision and improving an organisation's safety culture requires good team learning and dialogue. Team learning is about building high performance teams that can learn as a team and everyone can tap into each other's areas of competence to solve problems and achieve team goals. Senge (2006, p. 9) indicated that team learning starts with team dialogue or the ability to "suspend assumptions and enter into a genuine 'thinking together'". During an effective team dialogue, team members explore a complex issue, e.g. WSH-related issues, from different angles with the common aim of gaining clarity and fundamentally improving the system. This can only arise if the team has a shared vision and are able to let go of their egos and personal gains to focus on the "win" at the system level. The consequence of effective team dialogue is team learning, and if there are enough teams doing this in the organisation, organisational learning is achieved. Team learning is also about identifying possible team learning dysfunctions and actively preventing them from happening. In WSH management, the basic unit of WSH management is usually a team. Risk assessment (RA), incident investigation and many other WSH management activities are frequently conducted at the team level. If an RA team is not able to have open and meaningful discussions about hazards, it is a clear sign that the team and perhaps the organisation has a weak safety culture. Therefore, safety culture is closely related to team learning.

During direct interventions on safety culture, where dialogues are being conducted to discuss safety culture, participants of the dialogue sessions must suspend their assumptions, regard one another as colleagues, and the facilitator must be able to hold the context of the dialogue (Senge, 2006). After the dialogues, discussions must be conducted so that participants advocate possible visions and approaches that the team must deliberate, and then adopt. This process of dialogue (open exploration) and discussion (advocacy and decision making) is an important communication and consultation process in the direct intervention approach.

As can be observed, the concept of a learning organisation is closely related to safety culture, and organisations trying to improve safety culture can use the five disciplines proposed by Senge (2006) as a guiding framework for their interventions.

8.3 Change Management Models

8.3.1 Lewin's change model

One of the well-known change models is Lewin's change model (Cummings *et al.*, 2015), which states that change goes from *unfreezing* to *moving* and then *refreezing*. *Unfreezing* involves the removal of resistance to change through understanding the reason for resistance and providing information and education to the relevant stakeholders. *Moving* refers to the process of changing behaviours, attitudes and values through the establishment of new policies, structure and processes. *Refreezing* creates a new equilibrium by reinforcing the changes that were made and ensuring their sustainability. Prior to the *unfreezing* stage, the organisation must identify the need for change and the problems that can arise if the change is not implemented. Throughout the change process, there is a need for data to be collected about the current situation, the change process, and the situation after the change process has been completed. In addition, the team overseeing the change will have to develop frequent communication and coordination opportunities for stakeholders to give feedback and share their challenges and concerns.

8.3.2 Rider-elephant-path model

Heath and Heath (2010) gave an interesting analogy of the change process. According to them, each person has an emotional elephant and a rational rider. The rider riding the elephant is objective and rational, but may not be able to control the big elephant, which is focused on the short term and is guided by feelings. At the same time, if you want to influence the behaviour of the rider and his elephant, you will have to shape the path that they are trudging on. Based on this analogy, Heath and Heath (2010) suggested nine key change management actions — three for each of the components in the analogy (the rider, the elephant and the path). Even though the following discusses each of the components separately, all three components interact with each other and it is difficult to clearly distinguish the effects of each component. In addition, not all nine change management actions are relevant in all situations. Change leaders need to select a suitable combination of actions to encourage change.

8.3.2.1 Direct the rider

The Rider is a "thinker and planner" but has a tendency to "over-think" (decision or analysis paralysis) and over-focus on problems. To direct the rider towards the desired behaviour, a leader should find "bright spots" (positive examples) for followers to follow, script the critical actions very clearly, and make the desired outcome crystal clear.

Finding the bright spot is about looking for "insiders" who have used good or best practices to achieve desired results or top performance. An important note is that the manager should look for *practice* and not just *knowledge*. For example, many contractors know that it is important to avoid working-at-height (knowledge), but very few actually *practise* work methods that minimise working-at-height consciously. Furthermore, it is important to emphasise bright spots that arose from the "locals" (of a project/company) so that the "not invented here" problem will not arise, and workers will be more receptive of the good practice. It is also interesting to note that many of the good or best practices are very small scale and simple. Once the bright spots are identified, leaders must thoroughly investigate why the bright spots are different from the rest. The success factors must be identified so that it is possible to clone and reproduce as many bright spots as possible. In this way, the norm in the group changes. Cloning involves processes such as getting employees to shadow identified bright spots and getting bright spots to conduct sharing sessions.

Finding a bright spot is essentially a question of: "What's working and how can we do more of it?" It is solution-focused and not problem- focused. It is about scaling up solutions.

Since the rider can easily over-think, it is important for change leaders to narrow down the possible options, script the critical moves and direct the energy of workers towards the desired behaviour. Too many options delays change, and employees may choose the *status quo* (i.e. refuse to change) if it is too difficult to decide. Thus, resistance to change may be due to lack of clarity on the desired behaviour. For individuals to implement the change, there must be crystal clear guidelines on the specific desired behaviour. Because it is not possible to script every single move, managers must focus on the critical moves. For example, a company might want workers to *be safe* — but what is "being safe"? This is where organisations, through consultation with workers and supervisors, need to define the specific safe behaviours for each type of trade and task.

Safe behaviour can be defined through safe work procedures (SWPs), which were discussed in Chapter 4. Employers and principals are required to conduct RAs and derive SWPs based on the RAs. The SWPs have to document the specific risk controls that need to be implemented. Managers, supervisors and workers are expected to implement the risk controls, but if the SWPs were not clearly written and communicated, employees may not be able to implement the SWPs on the ground. The instructions for implementing the SWPs need to be written in a form that is easy for employees to understand.

Concurrently, organisations will need to make sure that the goals or desired outcomes are crystal clear so that the rider can focus his/her plans and decisions towards the goals. At the same time, having clear and desirable goals will also motivate the elephant.

8.3.2.2 Motivate the elephant

The elephant is the rider's biggest challenge. To motivate the elephant, find the feelings that influence people's behaviour (e.g. the joy of being with one's family), shrink the change into comfortable bite-sized or baby steps, and grow the people so that they feel a sense of identity and believe that they can change.

Heath and Heath (2010) cited Kotter and Cohen (2002) who observed that in most successful change efforts, the change process is not ANALYSE–THINK–CHANGE, but rather SEE–FEEL–CHANGE. In the ANALYSE–THINK–CHANGE model, a person analyses the situation, thinks of solutions and then changes to implement the solutions. The reality is that most of the time when we see the situation or something that grabs our attention, our emotions are stirred, and they influence our decisions and responses towards the situation. Thus, if the feeling that arises is not aligned with the intended change, change will not occur.

For example, if a campaign on "safe hands" is being implemented at a factory, even if many workers are invited to attend talks on how to protect their hands, because of the cognitive bias of over-confidence (discussed in Chapter 4), there is a tendency for the workers to feel that the likelihood of themselves being the victim of a hand-related accident is very small. In such a situation, the workers tend to ignore the safety measures promoted in the campaign. The campaign did not motivate the elephant because the messages did not stir any feelings nor help the workers relate to the situation. To motivate the elephant, it is more impactful to link the desired change or behaviours to positive emotions like happiness, camaraderie among workers, and recognition, rather than negative emotions like shame, the fear of being punished and the fear of losing one's family.

"Shrink[ing] the change" refers to cutting up a change into bite-size chunks so that it is not too daunting to implement. The elephant "hates doing things with no immediate payoff". Therefore, to evoke change, the first step or five-minutes are critical. If the employee is willing to start implementing the intended change, then there is a chance for the momentum to pick up so that the change might actually be implemented. The first task for effecting change should be simple and meaningful, and employees should be recognised for their initial effort.

Another approach to motivate the elephant is to grow the people through inspiration, so that they become bigger than the problem. The crux is the identity that people adopt. If each employee sees himself or herself as a safety champion or the workplace as an extension of the family or friends, then helping to spot hazards and unsafe behaviour becomes natural. Thus, the key challenge for change leaders is how to make the intended change a matter of identity and not merely a detached cost-benefit analysis. The key test is this, "Would an employee (who is being asked to change) aspire to be the kind of person who would make this change?" If the answer is yes, the elephant can be more easily motivated to initiate the change. If the answer is no, then there is a need to review if the change is suitable or if the perceived identity needs to be re-designed. The change leader needs to be mindful of the following identity questions when considering their change initiative, "Who am I? What kind of situation is this? What would someone like me do in this situation?"

Another difficulty that the elephant might face along the way is that change, especially in terms of organisational culture issues like safety culture, is a long and challenging path. How can one continue to motivate the elephant? Employees must be convinced that though there will be failures along the way, these expected failures do not end the mission. This mentality is an important aspect of the growth mindset, which praises effort rather than natural skill and believes in the ability of people to grow and adopt new skills along the way. Thus, a learning culture (as highlighted in Chapter 7), which includes the belief that people can change and improve, is critical to building safety culture. When incidents and accidents occur as a company embarks on a change journey to improve its safety culture and safety performance, the morale of employees may be affected, especially those directly involved in the incidents or implementation of change. This can derail the whole change effort. Thus, employees must be prepared to face failures and learn from them.

8.3.2.3 Shape the path

To shape the path, organisations must tweak the environment so that it is supportive of the desired behaviour, build habits through action triggers, checklists or other aids, and make use of the herd effect to spread the behaviour that the change process is hoping to build.

There is a natural tendency to blame individuals for their error, but as discussed in earlier chapters, many-a-times, the errors or unsafe behaviours were induced by the environment that they were in. For example, in the past, some Automated Teller Machines (ATMs) were designed to give you the money first before returning the card and this caused many people to forget to retrieve their ATM cards. A simple change in sequence, where the ATM card is returned before the money is given to the customer, resulted in a significant drop in the chances of customers forgetting to retrieve their ATM cards. The idea of tweaking the environment overlaps with the field of human factors and ergonomics, which design the environment and equipment to suit human characteristics, alleviate human limitations and minimise errors.

Another simple but effective approach to improve safety behaviour is to ensure that the work environment is neat and tidy. For example, if a construction site has poor housekeeping, where objects are placed near the edges of openings or buildings, and there are slipping hazards, it is a contravention of Reg 24 and 26 of the WSH (Construction) Regulations 2007 in Singapore. When workers work in an environment where WSH contraventions are the norm, they can easily form unsafe attitudes and there is a higher tendency to violate other WSH rules and standard operating procedures. On the other hand, if workers work in an environment where housekeeping is consistently practised, where equipment and materials are kept in the correct locations, the perceived norm would be that WSH rules and procedures are important and adhered to. This will lead to a change in the behaviour of individuals placed in this environment.

Shaping the path is also about creating habits that are automatic. To trigger these habits or routines, change leaders can develop action triggers such as posters, signage and hand signals, which can be incorporated into training. Workers can be trained on abbreviations like STOP (Stop, Think, Observe and Plan) during training. Complementary posters can be placed in the workplace to trigger the routines taught in training, which includes taking a pause during work to think about the possible hazards and controls in the current task and location, and to observe the task and location so that a plan can be put in place to ensure that the work can be conducted safely. These action-triggers need to be easy to understand and the actions required need to be carefully scripted as discussed earlier.

Another commonly used trigger is a checklist. A crane operator is required to conduct a pre-operation check according to a checklist. The checklist is a simple intervention, but it helps to ensure that the check is thorough and prevents over-confidence. However, to ensure that the checklist is used properly, the operator must understand its importance and supervisors must inspect the completed checklists every day and ask probing questions regarding the checklist. If the crane operators feel (i.e. the elephant in the operators) that the checklists are for novices, then the checklists will not be used even if they are useful. One way to minimise this is to relate the triggers and desired safe behaviours to positive identities like pilots, who use checklists diligently and professionally.

The last approach to shape the path is to rally the herd or to make safe behaviour contagious. This is focused on the behavioural norms of the target group the change leader is trying to change. Individuals tend to behave in accordance to the norm in a group. This has been supported by research such as Goh and Binte Sa'adon (2015), who found that the norm, as perceived by workers, has a strong influence on the intention and actual behaviour of a worker hooking on his safety harness.

The elephant is always looking to the herd for cues about how to behave. Workers with unsafe behaviours can be identified through inspections and behavioural observations. Once these workers are identified, supervisors and change leaders can get them to change by assigning buddies with safe behavioural habits to the worker to remind the worker to work safely at the beginning of and during shifts. The intention is to work through the smallest possible group to influence the individual's behaviour, to bring about positive cultural change. Concurrently, individuals identified for their safe behaviour will need opportunities to share their experience in implementing the desired change and supporting each other in the effort.

Heath and Heath's (2010) Rider-Elephant-Path model is a compilation of different research on change management. The model is simple and practical for change at individual and different organisational levels. Not all the nine approaches highlighted are always useful, and overlaps exist between the different approaches, so change leaders need to consider the situation and use the model selectively.

8.3.3 Senge's change model

Another useful literature on change management is Senge (1999), which highlighted three basic virtuous cycles that need to be in place for organisational change to be implemented and sustained (Figure 8.2):

1. Organisation's Safety Performance Results: The investment to build safety culture will take some time (delay) before capabilities in WSH management and risk controls can be improved. After a sustained period, the behaviours of people in the organisation will be improved and it will become a new norm for people to work safely. Since unsafe acts and unsafe

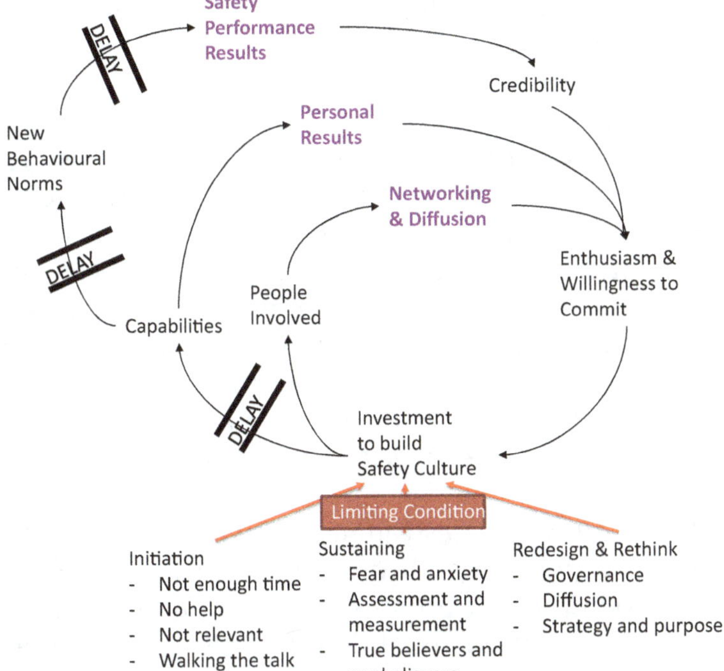

Figure 8.2   Virtuous cycles and limiting conditions; adapted from Senge (1999)

conditions will be reduced, managers should expect to see an improvement in WSH performance. The improvement will improve the credibility of the initial investments and interventions. This will then encourage more investment in the programme and spark off a virtuous cycle of improvement.

2. Networks of Committed People: The initial investment to build an organisation's safety culture will get people involved in improving WSH. The people that are directly involved will typically be more committed, but their initial numbers may be small. Getting these early supporters to network and diffuse their commitment and support will help to promote enthusiasm for the programme. This will encourage stronger commitment to the safety culture programmes and further investment of resources and time.

This virtuous cycle will sustain the programme and keep people engaged.

3. Personal Results: The initial investment to build safety culture will increase the organisation's WSH capabilities and after some time, the organisation should ensure that individuals see personal results from their improved WSH management capabilities. This means that the way that organisations measure and reward performance need to encourage good safety behaviour. Once individuals see positive consequences from their involvement in the safety culture programme, they will be more enthusiastic and more willing to commit to the programme. This will then develop into a virtuous cycle at the personal level.

All three virtuous cycles will help organisations to reap the benefits of their safety culture improvement programme. Managers trying to get organisations to change their safety culture or implement organisational changes will need the help of these virtuous cycles.

However, for the organisational change to be successful, it is critical for the organisation to remove barriers to the three virtuous cycles (see Figure 8.2). The barriers and solutions identified by Senge (1999) are briefly described in Table 8.1.

8.4 Change Management Plan

It is useful to have a change management plan with detailed documentation to guide and manage the organisational change process. An example can be found in the Queensland Government Chief Information Office (n.d.), which includes the following components:

1. Change identification: Type, reason, scope, current status, desired state
2. Change specification: policy, structure, processes, relationships, culture, people, information, cost, risk assessment

Table 8.1 Summary of the key change barriers and solutions; adapted from Senge (1999)

Barriers or Challenges	Possible Solutions
Initiation	
1. "We don't have time for this stuff" • Lack of control over one's time. People involved in change initiatives need enough flexibility to devote time to reflect and practise • People might be committed, but their lack of time creates an inability to meet the demands of the change initiative • People can end up frustrated and give up on the initiative • Especially true for pilot team members initiating change	• Integrate initiatives • Proper scheduling • Trust people to use their own time • Value unstructured time • Build productivity capability • No politics or games-playing and nonessential demands
2. "We have no help!" • Inadequate coaching, guidance and support for innovating group • No assistance to help build capacity to sustain change	• Invest in help — consultants, internal experts, guides, etc. • Creating capacity for coaching • Find a partner to confide in • Building coaching into line management • Treat seeking help as norm and positive
3. "This stuff isn't relevant!" • The absence of a clear, compelling business case for learning and change • Lack of commitment and enthusiasm	• Build awareness among key leaders • Explicitly raise questions about relevance in the pilot group

(Continued)

Table 8.1 (*Continued*)

Barriers or Challenges	Possible Solutions
Initiation	
	• Make more information available to pilot group members — connect to individual and job goals • Keep training linked tightly to business results. • Revisit relevance periodically
4. "They're not walking the talk!" • Lack of clarity and credibility of the management's aims and values • Lack of trust in management to carry out change	• Develop espoused aims and values that are credible • Demonstrate the values and commitment to aims through actions • Work with partners • Cultivate patience under pressure and with bosses • Develop a greater sense of organisational awareness (go to the ground) • Reflect on your beliefs about people and discuss values
Sustaining	
5. Fear and anxiety — Common questions: • Am I safe? • Am I adequate? • Can I trust myself and others to sustain the change?	• Start small and build momentum before confronting difficult issues • Avoid frontal assaults • Set an example of openness • Learn to see diversity as an asset • Use defects as opportunities for learning

Table 8.1   (*Continued*)

Barriers or Challenges	Possible Solutions
Sustaining	
	• Make sure participation in pilot groups and change initiatives is by choice, not by coercion • Remember, fear and anxiety are natural; forcing creates more anxiety
6. Assessment and measurement • Expectations of results • Delay in seeing results • Negative results due to change • Lack of measurement of results	• Appreciate time delays in change initiatives • Assign suitable metrics (not just cost or final outcome) • Recognise and appreciate progress
Redesign and Rethink	
7. True believers vs. nonbelievers (lack of understanding between groups) • Pilot group (believers) vs. those not in the pilot (nonbelievers) • Fear and anxiety arising among those not involved in the change • Lack of engagement between groups • Arrogance among pilot group members who think they are always right • Pilot group might feel under-appreciated for their efforts	• Become "bicultural" — incubators and mainstream become distinct groups • Mentoring of line leaders • Build ability to engage so as to spread change initiative beyond pilot group • Cultivate reflective openness • Respect people's inhibitions about personal change • You don't have to convince people (mutual understanding) • Remove jargons and keep language simple • Lay foundation of transcendent values (tolerance and flexibility)

(*Continued*)

Table 8.1 (*Continued*)

Barriers or Challenges	Possible Solutions
Redesign and Rethink	
8. Governance ("Who is in charge of this stuff?") • How to institutionalise the change initiatives and align them with existing structures • Balance between autonomy for change and corporate control	• Pay attention to boundaries and be strategic when crossing them • Articulate the case for change in terms of business results • Make executive leaders' priorities part of your creative thinking • Experiment with cross-functional, cross-boundary teams, if you can get them sponsored by the hierarchy • Begin at the beginning: with governing ideas • Deploy new rules and regulations judiciously • Never underestimate the power of small changes in complex situations if they are the "right" changes • Be prepared for a long journey and don't embark alone
9. Diffusion ("Reinventing the wheel") • Not learning from pilot groups' experience • Innovations by others being dismissed or regarded as not applicable to them • Lack of interest	• Increase quality and number of coaches • Increase permeability of intra organisational boundaries • Improve information infrastructure — share innovations and changes

Table 8.1   (*Continued*)

Barriers or Challenges	Possible Solutions
Redesign and Rethink	
• Impatience in repeating change initiatives • Arrogance — Assume that there is nothing new to learn	• Build a learning culture and communities of practice • Include research as part of executive accountability
10. Strategy and Purpose • Change can create questions about the existing strategy and purpose • What do we really want to create? • How will it contribute to others? • New ideas on strategy and purpose may not be accepted	• Use scenario thinking to investigate blind spots, signals of unexpected events and organisational purpose • Engage people in organisational strategy and purpose • Expose and test assumptions behind strategy

3. Change methodology: Stakeholder analysis (target groups, advocates, early adopters, resistance and drivers)
4. Implementation strategies: Action plan, schedules, communication plan, training plan, IT/business systems plan, resistance management plan

8.5 Case Study: Alcoa

This case study is adapted from Duhigg (2014) (Chapter 4), which describes how Alcoa, formerly known as The Aluminum Company of America, successfully established a culture of excellence by focusing on safety as a "keystone habit". Alcoa is a global industrial leader that has businesses in different industries. In 1987, Paul O'Neill, was appointed as the new CEO. He was a former government bureaucrat and was relatively unknown in the Wall Street community. O'Neill made a shocking announcement in his first shareholder meeting. He indicated that he intended to make Alcoa the safest company in America and he wanted to go for zero injuries. This was unprecedented and investors in the room were shocked.

O'Neill believes that bringing down the injury rates would mean that individuals in the company have agreed to devote themselves to creating a habit of excellence. In this way, safety becomes an indicator of how good the company is in changing shared habits. O'Neill wanted to attack one habit and then let the effect of the change ripple through the organisation. This is what Duhigg (2014) defined as a "keystone habit", a habit that has the power to start a chain reaction to influence how all employees behave and communicate in the organisation. These keystone habits are a focus and leverage for managers to initiate fundamental changes in organisational culture. Instead of framing safety culture as a subset of organisational culture, safety culture becomes a trigger for change in organisational culture.

O'Neill required managers to understand why injuries happen. In the process of investigating incidents, managers had to understand how the manufacturing process went wrong so as to understand and rectify the process problems. Workers needed to be educated about quality control, efficient work processes, how to make processes less error-prone, etc. The entire campaign centred around the idea that correct work is safe work and vice versa.

O'Neill's commitment was put to the test when a fatality occurred. An extrusion press (a metal working machine) had stopped operating; a worker jumped over the yellow safety wall surrounding the press and walked across the pit to remove a piece of aluminium jammed into the hinge of a swinging six-foot arm. When the jammed aluminium scrap was removed by the young man, the machine jumped back into motion, and the mechanical arm swung and struck the worker on his head, crushing his skull and killing him. The worker was a young employee who had just joined the company a few weeks ago. The worker was eager to take up the job because his wife was pregnant, and they needed the health care the job offered.

O'Neill was informed about the accident in the middle of the night and within 14 hours, ordered all the plant's executives and Alcoa's top officers to its headquarters to attend an emergency meeting. They reviewed the evidence and analysed the accident in detail. They identified several causes including:

- two managers saw the worker jump over the barrier but failed to stop him
- a training programme that failed to emphasise to the man that he would not be blamed for a breakdown
- lack of instructions that he should find a manager before attempting a repair
- absence of sensors to automatically shut down the machine when someone stepped into the pit

The CEO said, "We killed this man... It's my failure of leadership... And it's the failure of all of you in the chain of command." This is a very strong statement, but more indicative of his commitment were his subsequent actions. O'Neill gave his home telephone number to workers on the ground, asking them to give him a call whenever there were safety issues that their managers did not follow up on. According to Duhigg (2014), workers started to call, and they were giving O'Neill great ideas to improve work in general.

One example is when a worker made a suggestion to improve the work process at an aluminium manufacturing plant that manufactured siding for houses. The plant had engaged consultants to choose shades of paint to anticipate the colours that customers would want, but despite the hefty consultant fees, the problem remained unsolved. The worker suggested that all the painting machines be grouped together, so that paint pigments could be switched out faster, allowing them to be nimbler in responding to changes in customers' colour preferences. This simple suggestion resulted in profits on aluminium siding within a year. Interestingly, the worker had had the idea for a decade, but he had not told his management. Instead, it was only when O'Neill started focusing on improving safety, that the communication channel was opened — and that was when the employee thought that his suggestion might be useful, and finally decided to raise it. The focus on improving safety culture created a climate for innovation and willingness to change.

By 2000, when O'Neill retired, the annual net income had increased by five times as compared to when O'Neill initially started. Market capitalisation had risen by US$27 billion and Alcoa had become one of the safest companies in the world. More specifically, prior to 1987, the Alcoa plant had at least one accident per week, but after O'Neill took over as CEO, there were plants that had not had accidents that caused more than one day of work to be lost. The company's worker injury rate was one-twentieth of the US average.

This case study shows that improving safety culture is possible, and Alcoa is a bright spot that should be cloned. The importance of change leaders cannot be overstated, and it is obvious that the change was difficult, but the benefits go beyond WSH.

8.6 Conclusions

Improving safety culture is not an easy task. Managers need to structure the change systematically and deliberately. This chapter highlighted several approaches for improving safety culture. Safety culture intervention is a specific type of organisational change; hence it is useful to understand a range of organisational change management models. When managers and change leaders try to improve safety culture, the organisation needs to learn the new culture. Therefore, Senge's (2006) model of learning organisation, which involves the disciplines of systems thinking, personal mastery, mental model, shared vision, and team learning and dialogue, is a possible guiding framework for the change initiative. This chapter also introduced other change management models, including Lewin's model, Heath and Heath's (2010) Rider-Elephant-Path model, and Senge's (1999) virtuous cycles and barriers. A template for a change management plan was also briefly discussed. Lastly, Alcoa's successful change initiative was presented as a bright spot of how organisational excellence was improved through a safety improvement campaign.

Review Questions

1. What are the three general approaches to improving safety culture?
2. Compare the five disciplines of a learning organisation with Reason's model of safety culture. What are the similarities and differences?
3. Compare Heath and Heath's (2010) model of change with Lewin's model. What are the similarities and differences?
4. Describe the three virtuous cycles needed to make sustainable changes in organisations.
5. According to Senge (1999), what are the 10 types of barriers to change?
6. Apply the Rider-Elephant-Path model to Alcoa's case study. What were the approaches used to implement the change?

References

Cummings, S., Bridgman, T., and Brown, K. G. (2015). "Unfreezing change as three steps: Rethinking Kurt Lewin's legacy for change management". *Human Relations, 69*(1), 33–60. doi: ttps://doi.org/10.1177/0018726715577707

Duhigg, C. (2014). *The Power of Habit: Why we do what we do in Life and Business.* New York: Random House.

Goh, Y. M., and Binte Sa'adon, N. F. (2015). "Cognitive factors influencing safety behavior at height: A multimethod exploratory study." *Journal of Construction Engineering and Management*, 141(6).

Heath, C., and Heath, D. (2010). *Switch: How to Change things when Change is Hard*. New York: Broadway Books.

Kotter, J. P. and Cohen, D. S. (2002). *The Heart of Change: Real-life Stories of how People Change their Organizations.* Boston, Mass: Harvard Business School Press.

Queensland Government Chief Information Office (n.d.). "Change Management Plan Workbook and Template." http://www.nrm.wa.gov.au/media/10528/change_management_plan_workbook_and_template.pdf. (accessed March 12, 2018).

Senge, P. (1999). *The Dance of Change: The Challenges of Sustaining Momentum in Learning Organizations: A Fifth Discipline Resource.* London: Nicholas Brealey Pub.

Senge, P. M. (2006). *The Fifth Discipline: The Art and Practice of the Learning Organization* (First, revised and updated.). New York: Currency.

CHAPTER 9

Overview of Workplace Safety and Health Legislations

9.1 Introduction

In most countries, workplace safety and health (WSH) management is regulated by a suite of regulations. As part of any WSH management system, the organisation needs to have procedures to identify and access relevant WSH legislations and ensure compliance to the regulations (see Clause 6.1.3 of ISO 45001:2018). This chapter provides a brief overview of the Singapore Workplace Safety and Health Act (WSHA) and its subsidiary legislations, which have similar requirements as the WSH legislations in other countries. This chapter relies heavily on the content provided on the Ministry of Manpower (MOM) website and Singapore Statutes Online.

Table 9.1 contains the list of legislations under the WSHA (Cap. 354A). The WSHA was discussed in Chapter 1. In addition, some of the subsidiary regulations, such as the Workplace Safety and Health Design for Safety Regulations 2015 (S428/2015) and the Workplace Safety and Health (Risk Management) Regulations (Cap. 354A, RG 8), were also discussed in the relevant chapters. Other details of the WSHA and selected regulations will be briefly discussed in this chapter.

An important point to note is that even though there could be other regulations that cover specific hazards, e.g. structural safety is also covered under the Building Control Act, the requirements in the WSHA and its subsidiary regulations cannot

Table 9.1 List of WSHA-related legislations in Singapore

#	Title	Number
1	Workplace Safety and Health Act	Cap. 354A
2	Factories (Registration and Other Services — Fees and Forms) Regulations	RG 5
3	Factories (Safety Training Courses) Order	Cap. 354A/104, O 12
4	Factories (Work of Engineering Construction) Order	Cap. 354A/ 104, O 6
5	Workplace Safety and Health (Abrasive Blasting) Regulations 2008	S 607/2008
6	Workplace Safety and Health (Asbestos) Regulations 2014	S 337/2014
7	Workplace Safety and Health (Composition of Offences) Regulations	RG 6/2007
8	Workplace Safety and Health (Confined Spaces) Regulations 2009	S 462/2009
9	Workplace Safety and Health (Construction) Regulations 2007	S 663/2007
10	Workplace Safety and Health (Design for Safety) Regulations 2015	S 428/2015
11	Workplace Safety and Health (Exemption) Order	O 1
12	Workplace Safety and Health (Explosive Powered Tools) Regulations 2009	S 325/2009
13	Workplace Safety and Health (First-Aid) Regulations	RG 4
14	Workplace Safety and Health (General Provisions) Regulations	RG 1
15	Workplace Safety and Health (Incident Reporting) Regulations	RG 3
16	Workplace Safety and Health (Major Hazard Installations) Regulations 2017	S 202/2017

Table 9.1   (*Continued*)

#	Title	Number
17	Workplace Safety and Health (Medical Examinations) Regulations 2011	S 516/2011
18	Workplace Safety and Health (Noise) Regulations 2011	S 424/2011
19	Workplace Safety and Health (Offences and Penalties) (Subsidiary Legislation under Section 66(14)) Regulations	RG 5
20	Workplace Safety and Health (Operation of Cranes) Regulations 2011	S 515/2011
21	Workplace Safety and Health (Registration of Factories) Regulations 2008	S 501/2008
22	Workplace Safety and Health (Risk Management) Regulations	RG 8
23	Workplace Safety and Health (Safety and Health Management System and Auditing) Regulations 2009	S 607/2009
24	Workplace Safety and Health (Scaffolds) Regulations 2011	S 518/2011
25	Workplace Safety and Health (Shipbuilding and Ship-Repairing) Regulations 2008	S 270/2008
26	Workplace Safety and Health (Transitional Provision) Regulations	RG 7
27	Workplace Safety and Health (Work at Heights) Regulations 2013	S 223/2013
28	Workplace Safety and Health (Workplace Safety and Health Committees) Regulations 2008	S 355/2008
29	Workplace Safety and Health (Workplace Safety and Health Officers) Regulations	RG 9
30	Workplace Safety and Health (Workplaces Subject to Act) Order 2007	S 72/2007

be automatically assumed to be satisfied when those other regulations are complied with.

9.2 Workplace Safety and Health Act (1st Jan 2018 Version)

The key duties of the different duty holders were discussed in Chapter 1. The following will provide additional information highlighted in the WSHA.

9.2.1 Difference between workplace and factory

The WSHA provided a distinction between a "workplace" and a "factory". A "workplace" means any premises where a person is at work or is to work, for the time being works, or customarily works, and includes factories. Some of these workplaces are further classified as a factory. A factory is a term carried over from the repealed Factories Act, which was the main legislation regulating industrial safety and health before the WSHA. A factory is any premises in which any of the following is carried out:

(a) The making of any article or part of any article.
(b) The alteration, repair, ornamentation, finishing, cleaning or washing of any article.
(c) Breaking up or demolishing any article.
(d) Adapting any article for sale.

As described in the Fourth Schedule of the WSHA, examples of factories include manufacturing plants, car-servicing workshops, shipyards and construction worksites. These workplaces are considered higher risk and are subjected to additional regulations, e.g. periodic audits and a mandatory WSH committee.

9.2.2 Protecting employees

The WSHA Section 18 protects employees by stating that employers shall not "(1)(a) deduct, or allow to be deducted, from the sum contracted to be paid by him to any employee of his; or (b) receive, or allow any agent of his to receive, any payment from any employee of his, in respect of anything to be done or provided by him in accordance with this Act in order to ensure the safety, health or welfare of any of his employees at work." Anecdotally, there are practices in the industry where employers (including principals) deduct the pay of employees or ask for payment for personal protective equipment (PPE) that should have been provided to the workers as required by WSHA. In addition, Section 18 also makes it illegal for employers to dismiss or threaten to dismiss an employee if the employee whistle-blows on the employer for WSHA-related contraventions, or for performing his duties as a member of the WSH committee.

9.2.3 Powers of commissioner

The WSHA gives the Commissioner for WSH, who is typically the Director of the Division of Occupational Safety and Health in MOM, powers to issue stop work orders (to stop the work at a workplace) and remedial orders (to improve the conditions in a workplace). Failure to comply with a remedial order can result in a fine not exceeding S$50,000 or imprisonment for a term not exceeding 12 months or both. In the case of a continuing offence, the duty holder may have to pay a further fine not exceeding S$5,000 for every day or part thereof during which the offence continues after conviction. On the other hand, failure to comply with a stop work order can result in a fine not exceeding S$500,000 or imprisonment for a term not exceeding 12 months or both. In the case of a continuing offence, the duty holder may have to pay a further fine not exceeding S$20,000 for every day or part thereof during which the offence continues after conviction.

The Commissioner also has the power to enter a workplace for the purpose of inspection and investigation. The WSHA also made provisions to protect accident-related evidence, where any alteration or addition to machinery, equipment, etc. related to an accident and dangerous occurrence is an offence.

Section 27A of the WSHA is a proactive amendment of the WSHA in 2018. The section allows the Commissioner to prepare and publish a learning report on any accident, dangerous occurrence or occupational disease in a workplace that is still being investigated. Prior to this section, it was difficult for MOM to disseminate important WSH information gained from investigations prior to prosecution because the information released may influence the outcome of the prosecution. The learning report may:

(a) contain an account of the accident, dangerous occurrence or occupational disease;
(b) specify the cause or causes of, and circumstances or factors leading to, the accident, dangerous occurrence or occupational disease insofar as they may be ascertained;
(c) contain an opinion by a person with technical or specialised knowledge of the machinery, equipment, plant, article, process, substance, work or workplace involved in the accident, dangerous occurrence or occupational disease;
(d) contain a warning of any danger or risk to the safety and health of persons at work or persons who may be affected by any undertaking carried on in the workplace;
(e) contain any recommendation to prevent or minimise the recurrence of any similar accident, dangerous occurrence or occupational disease in a workplace; and
(f) contain any other matter that the Commissioner considers relevant, taking into account the sole objective mentioned in subsection (2), which is "to prevent or minimise the recurrence of any accident, dangerous occurrence or occupational disease in a workplace, and not to apportion blame or liability."

The learning report is not admissible in any legal proceedings. This is aligned with the approach used by the US Chemical

Safety Board (CSB), where the investigation report written by CSB is not admissible in court and the US Occupational Safety and Health Administration (OSHA) conducts an independent investigation for its prosecution purposes.

9.2.4 Safety and health management arrangements

The WSHA Part VII specifies requirements for safety and health management arrangements. These requirements pertain to WSH Officers (WSHOs) and co-ordinators, WSH committees, WSH auditors, safety and health training courses and the need for the Commissioner's approval for people to perform statutory roles (authorised examiner for dangerous machines, WSHO, WSH co-ordinator, WSH auditor and accredited training provider). The WSHA only provided the framework to regulate each of these areas of safety and health management arrangements. The details of the actual requirements are usually found in relevant subsidiary regulations and government gazettes.

9.2.5 Workplace safety and health council and approved code of practice

The WSHA also provided for the establishment of the WSH Council. According to Section 40A, the functions of the WSH Council are:

(a) to develop or facilitate the development of acceptable practices relating to safety, health and welfare at work;
(b) to promote the adoption of acceptable practices relating to safety, health and welfare at work;
(c) to devise, organise and implement programmes and other activities for or related to providing support, assistance or advice to any person or organisation in preserving, improving and promoting safety, health and welfare at work;
(d) to facilitate and promote the development and upgrading of competencies, skills and expertise of the workforce relating to safety, health and welfare at work;

(e) to research into any matter relating to safety, health and welfare at work;

(f) to grant prizes and scholarships, and to establish and subsidise lectureships in universities and other educational institutions in subjects relating to safety, health and welfare at work;

(g) to provide practical guidance with respect to the requirements of this Act relating to safety, health and welfare at work; and

(h) to do all the things that it is authorised or required to do under the WSHA.

Sections 40B and 40C then stipulated the role of the codes of practice as issued or approved by the WSH Council. The codes of practice approved by the WSH Council are frequently Singapore Standards published by the Standards, Productivity and Innovation Board under Section 7(2)(h) of the Standards, Productivity and Innovation Board Act (Cap. 303A). Essentially, the approved code of practice (ACoP), is a non-mandatory benchmark for the court to determine whether a measure is reasonable and practicable. In the absence of a suitable ACoP, relevant Singapore Standards, not approved by the WSH Council and industry guidelines, can also be used to establish standards that are reasonable and practicable.

9.2.6 Schedules

The First Schedule provided the specific types of dangerous occurrences which must be reported to MOM. They include bursting failures of machinery, crane collapses, significant explosions or fires, significant electrical short-circuits or failures, explosions or failures of a steam boiler's structure(s), failures or collapses of formwork or its supports, scaffold (15 metres in height) collapses of suspended or hanging scaffolding above two metres, and accidental seepage or entry of seawater into a dry dock.

The Second Schedule identified 35 occupational diseases that must be reported to the Manpower Ministry:

1. Aniline poisoning	13. Diseases caused by excessive heat	25. Organophosphate poisoning
2. Anthrax	14. Hydrogen Sulphide poisoning	26. Phosphorus poisoning
3. Arsenical poisoning	15. Lead poisoning	27. Poisoning by benzene or a homologue of benzene
4. Asbestosis	16. Leptospirosis	28. Poisoning by carbon monoxide gas
5. Barotrauma	17. Liver angiosarcoma	29. Poisoning by carbon disulphide
6. Beryllium poisoning	18. Manganese poisoning	30. Poisoning by oxides of nitrogen
7. Byssinosis	19. Mercurial poisoning	31. Poisoning from halogen derivatives of hydrocarbon compounds
8. Cadmium poisoning	20. Mesothelioma	32. Musculoskeletal disorders of the upper limb
9. Carbamate poisoning	21. Noise-induced deafness	33. Silicosis
10. Compressed air illness or its sequelae, including dysbaric osteonecrosis	22. Occupational asthma	34. Toxic anaemia
11. Cyanide poisoning	23. Occupational skin cancers	35. Toxic hepatitis.
12. Diseases caused by ionising	24. Occupational skin diseases	

The Third Schedule provided a list of activities classified as Work of Engineering Construction, i.e. construction work, while the Fourth Schedule identified 19 categories of workplaces defined as factories. The Fifth Schedule listed 11 types of machinery and equipment (e.g. scaffolds, cranes, forklifts, bar benders and welding machines) and 17 types of hazardous substances (e.g. corrosive, flammable, explosive, oxidising, toxic and carcinogenic substances) that require their manufacturers and suppliers to ensure compliance with Section 16 of WSHA, which makes it compulsory to:

(a) Provide information on health hazards and how to safely use the machinery, equipment or hazardous substance.
(b) Examine and test the machinery, equipment or hazardous substance to ensure that it is safe for use.
(c) Provide results of any examinations or tests of the machinery, equipment or hazardous substances.

The persons who erect, install or modify the machinery and equipment identified in the Fifth Schedule are required under Section 17 to ensure that the method of work is in accordance to the manufacturer's instruction and that the machinery or equipment is safe for use after installation, erection or modification.

9.3 Workplace Safety and Health (Asbestos) Regulations 2014

Asbestos is a hazardous substance that can cause lung diseases such as asbestosis (scarring and fibrosis of the lung tissues), mesothelioma (a cancer of the chest and abdominal lining) and lung cancer (Workplace Safety and Health Council, 2017). The substance is not a specific mineral, but a group of fibrous silicate minerals (Tranter, 2004) that was commonly used for insulation, fire protection and acoustic purposes. The Workplace Safety and Health (Asbestos) Regulations 2014 defines asbestos as crocidolite, actinolite, anthophyllite, amosite, tremolite, chrysotile, amphiboles or a mixture containing any such minerals. The material has been banned in Singapore since 1989, but it can still be found in the thermal insulation of pipes in vessels, plants and furnaces, and buildings constructed before 1989. The National Environment Agency (NEA) and MOM regulate the use and disposal of asbestos in Singapore. MOM focuses on the protection of people at work, and the key legal instrument is the Workplace Safety and Health (Asbestos) Regulations 2014 ("Asbestos Reg").

The Asbestos Reg contains six parts and a Schedule. The Schedule defines the material, substance, product or article classified as "specified material" that requires special attention and measures under the Asbestos Reg. The Schedule listed the following "specified material":

1.   Cable penetration insulation
2.   Fire protection board, panel, wall and door
3.   Gasket
4.   Refractory lining
5.   Sprayed insulation
6.   Thermal insulation of pipe, boiler, pressure vessel and process vessel.

For any work that involves the above specified material that is likely to contain asbestos, or any demolition, alteration, addition or repair of a building built before 1 Jan 1991, the employer or principal must engage a competent person to assess if the specified material contains asbestos. The competent person will then conduct a survey to collect suitable samples and send it for testing by an approved testing body. A survey report must then be provided after the test with regards to the presence of asbestos.

If asbestos is present in the specified material, then the Occupier must ensure that the asbestos removal work is conducted by an Approved Asbestos Removal Contractor (AARC) (a list of AARCs can be found on the MOM website). Work involving asbestos must be conducted by workers who are trained to understand the hazards of asbestos and the safe way to handle the material, including training on the use of respiratory protective equipment. These workers will also be subjected to medical examinations stipulated in the Workplace Safety and Health (Medical Examinations) Regulations 2011. The AARC must engage a competent person to develop an asbestos-removal plan of work and supervise the asbestos-removal work. The AARC must also ensure the implementation of the plan. MOM must be notified of all asbestos removal work seven days prior to the commencement of the removal work.

9.4 Workplace Safety and Health (Confined Spaces) Regulations 2009

According to the Workplace Safety and Health (Confined Spaces) Regulations 2009 ("Confined Spaces Reg"), a confined space is defined as any chamber, tank, man-hole, vat, silo, pit, pipe, flue or other enclosed space, in which:

(a) dangerous gases, vapours or fumes are liable to be present to such an extent as to involve a risk of fire or explosion, or persons being overcome thereby;
(b) the supply of air is inadequate, or is likely to be reduced to be inadequate, for sustaining life; or
(c) there is a risk of engulfment by material.

Confined spaces are dangerous because they may not be easily identified by workers. There have been accidents where multiple workers were killed because when the first worker become unconscious in the confined space, co-workers entering the confined space to rescue the affected worker also become affected by the harmful gases or substances in the space. The Confined Spaces Reg stipulates the requirement for the Occupier to record the description and location of all fixed and stationary confined spaces and inform the relevant people of the hazards of these identified confined spaces. The Occupier must also ensure that the access and egress, entrance, lighting and ventilation of the confined space(s) are safe and suitable for the confined space(s).

The Confined Spaces Reg stipulates a permit-to-work system where a competent authorised manager and a competent confined space safety assessor are appointed. The confined space entry permit must specify the following:

(a) the description and location of the confined space;
(b) the purpose of entry into the confined space;
(c) the results of the gas testing of the atmosphere of the confined space; and
(d) its period of validity.

The Occupier is to evaluate if the confined space entry is indeed necessary. If the entry is necessary, the Occupier must ensure that the permit-to-work system is implemented. If the person entering the confined space is wearing a suitable breathing apparatus, authorised by the authorised manager to enter the confined space, and "where reasonably practicable, is wearing a safety harness with a rope securely attached and there is a confined space attendant keeping watch outside the confined space who is provided with the means to pull such a person out of the confined space in an emergency", then the confined space entry permit is not necessary.

The application for the confined space entry permit is to be made by the supervisor of the person entering the confined space. The submitted application form will have to contain the safety measures to protect the worker(s) entering the confined space. These safety measures will have to be assessed by the confined space assessor and then approved by the authorised manager before the workers can enter the confined space. The entry permit must be displayed at the entrance of the confined space.

One of the duties of the confined space entry assessor is to test the atmosphere of the confined space for oxygen content (19.5% to 23.5% by volume), level of flammable gases or vapour (less than 10% of its lower explosive limit), and the presence of toxic gas or vapour (that it does not exceed the permissible exposure levels as specified in the First Schedule to the Workplace Safety and Health (General Provisions) Regulations.

The gas meter used in the testing must be properly calibrated. The confined space safety assessor will have to conduct periodic testing of the atmosphere during the conduct of the work. In addition, if there is more than one person present in the confined space, at least one of them must have a suitable gas detector that continuously monitors the atmosphere in the confined space. In the event that a hazardous atmosphere is detected, all personnel must vacate the confined space. A re-evaluation of the confined space must be conducted, and a new permit will have to be issued before any worker can enter the confined space.

Another key hazard of a confined space is "incompatible work", which means work which is carried out at or in the vicinity of any work carried out in the confined space and which is likely to pose a risk to the safety and health of persons present in the confined space. A set of incompatible work includes hot work (e.g. welding, hot-cutting and grinding) and spray painting in confined spaces. The work may be in adjacent confined spaces, but when the incompatible substances come into contact a major fire and explosion can occur. Thus, the Confined Spaces Reg requires anyone who is aware of incompatible work to immediately report it to their supervisor, WSH personnel and authorised manager.

9.5 Workplace Safety and Health (Construction) Regulations 2007

The Workplace Safety and Health (Construction) Regulations 2007 ("Con Reg") is one of the core WSH regulations for the construction industry. It contains 16 parts and 142 Regulations.

In addition to the Workplace Safety and Health (Safety and Health Management System and Auditing) Regulations 2009, the Con Reg requires the Occupier to convene site coordination meetings to coordinate site activities so as to ensure the safety, health and welfare of persons at work in the worksite. The site coordination meeting has to be chaired by the "project manager", which means the person who is stationed at a worksite and who has overall control of all the works carried out in the worksite, and includes any competent person appointed by the occupier of the worksite in the event that the project manager is unable to perform his duties under the Con Reg. In addition, the Occupier of a project with a contract sum of less than $10 million will have to appoint a WSH co-ordinator to assist the occupier in identifying unsafe conditions or unsafe work practices, recommend to the occupier reasonably practicable measures to remedy the unsafe conditions or unsafe work practices, and assist the occupier in implementing the recommended reasonably practicable measures. This role is similar to that of a WSHO, but the level of training of a WSH co-ordinator is of a lower level than a WSHO. The Con Reg also requires suitable safety and health training for all workers and supervisors.

Similar to the Confined Spaces Reg, the Con Reg also requires a permit-to-work system for the following high-risk construction activities:

(a) demolition work;
(b) excavation and trenching work in a tunnel or hole in the ground exceeding a depth of 1.5 metres;
(c) lifting operations involving a tower, mobile or crawler crane;
(d) piling work; and
(e) tunnelling work.

The remaining portions of the Con Reg cover different hazards, such as structural safety and stability, storage and stacking of materials and equipment, falling objects, slipping hazards, protruding hazards, personal protective equipment, electrical safety, cantilevered platforms, formwork, demolition, excavation and tunnelling, piling, crane and employee's lifts.

9.6 Workplace Safety and Health (General Provisions) Regulations

The Workplace Safety and Health (General Provisions) Regulations ("GP Reg") covers all workplaces. It covers general provisions relating to health, safety and welfare. The hazards identified under health issues include: infectious agents and biohazardous material, overcrowding, ventilation, lighting, drainage (to prevent wet floors), sanitary conveniences, vibration and excessive heat or cold and harmful radiations.

The range of safety hazards covered in the GP Reg is much wider. Several of the Regulations cover machinery hazards. For example, Regulations 11 stipulates the requirements for occupier to ensure secure fencing (or guarding) for prime movers (e.g. electric motor and internal combustion engines) and connecting flywheels and moving parts. Regulations 12 further indicates that fencing is not necessary only if the dangerous parts of the machinery are made safe (i.e. no person can be harmed) because of its position, construction or other means. Another requirement is the need for an emergency cut off that can efficiently cut off the power to the machinery. Regulations 13 requires safety measures to be implemented to assure the safety of maintenance workers who may need to remove the machinery fencing while the machinery is operating. The worker must be at least 18 years old. He must have been trained for maintenance or repair jobs and he must be wearing suitable clothing that does not have loose ends.

Other safety requirements include electrical installation and equipment (prevention of electrical shock), and requirements to ensure fencing and other safeguards are properly constructed, used and maintained. The occupier also needs to set up a set of lock-out procedures to prevent accidental activation or energising during the inspection, cleaning, repair or maintenance of any plant, machinery, equipment or electrical installation in the workplace. The Regulations also cover the danger of a person falling into the tank, structure, sump or pit containing scalding, burning, corrosive or toxic liquid. Self-acting machines (e.g. robotic arms) are required to be made safe by, for example, ensuring no person can be exposed to the machine in operation, and having warning signs.

Regulations 19 to 21 (hoists and lifts, lifting gears, lifting appliances and lifting machines) are related to the role of an Authorised Examiner (AE), who is essentially a special class of Professional Engineers authorised by MOM to conduct the inspection and certification of dangerous lifting-related machineries. The occupiers and AEs play important roles in preventing accidents like lifted objects falling from height and failure of hoists.

Even though there is a Workplace Safety and Health (Work at Heights) Regulations 2013 ("WAH Regs"), the GP Regs contains several requirements on fall from height hazards. Regulations 23 imposes a duty on the occupier to ensure that all openings in the floors of the workplace are securely covered or fenced unless the fencing is impracticable. Regulations 23 requires employers to provide "secure foothold(s) and handhold(s)", as far as is reasonably practicable, for any worker working at 2 metres or higher or if the worker can fall into any substance that can cause drowning or asphyxiation. If secure footholds and handholds are not reasonably practicable then the employer must provide other suitable means such as safety harnesses or safety belts (i.e. fall arrest or travel restraint systems). It was stipulated that the anchorage for the harnesses or safety belts must not be lower than the level of the working position of the worker.

Other safety hazards covered included storage of goods, explosives, flammable dust, gas, vapours or substances, pressure vessels (e.g. steam boiler, steam receiver, air receiver and refrigerating plant, which require AE inspection and certification) and fire.

Part IV covers toxic dust, fumes or other contaminants, permissible exposure levels of toxic substances, hazardous substances, warning labels and safety data sheets. It is noted that the occupier must ensure that any hazardous substance found in a workplace must have a safety data sheet. The First Schedule of the GP Reg shows the permissible exposure levels (PELs) for a wide range of hazardous hazards.

9.7 Workplace Safety and Health (Incident Reporting) Regulations

Under the Workplace Safety and Health (Incident Reporting) Regulations ("IR Reg") the employer is responsible for notifying the Commissioner of Workplace Safety and Health (in practice this will be through MOM) of any fatal accidents involving its employees as soon as is reasonably practicable. The occupier will have to report the death of non-workers and the self-employed to the Commissioner. The report must be made within 10 days of the accident. If a dangerous occurrence occurs (see First Schedule of the WSHA), the occupier shall, as soon as is reasonably practicable, notify the Commissioner of the occurrence. This must be done not later than 10 days after the occurrence.

As of 31 August 2020, in the event of an injury, the employer must submit a report to the Commissioner if the employee "is certified by a registered medical practitioner or registered dentist to be unfit for work, or to require hospitalisation or to be placed on light duties, on account of the accident", within 10 days after the date the employer first has notice of the accident. If the injury was sustained by a person who was not working or is self-employed, the occupier will have to notify the Commissioner.

In terms of occupational diseases listed in the Second Schedule in the WSHA, the employer and registered medical practitioner shall each submit a report to the Commissioner within 10 days of the diagnosis.

The notification reports are legal documents and employers and occupiers must maintain a copy of the reports for at least three years.

9.8 Workplace Safety and Health (Operation of Cranes) Regulations 2011

Cranes are dangerous machines that can kill many people during an accident. Thus, the Workplace Safety and Health (Operation of cranes) Regulations 2011 ("OOC Reg") identified a list of duties and requirements for the responsible person, which is frequently the occupier. One of the key administrative controls is the lifting plan. The lifting plan is described in more detail in the Code of Practice on Safe Lifting Operations in the Workplaces (Workplace Safety and Health Council, 2014). Cranes exceeding 5 tonnes and tower cranes can only be operated by a registered crane operator. Cranes not exceeding 5 tonnes (e.g. a mini crane or lorry crane) can only be operated by a competent operator, but the operator need not be registered with the Commissioner for WSH.

A registered crane operator is required to:

(a) conduct a pre-start inspection on the crane;
(b) ascertain the ground conditions and report to the lifting supervisor if he feels that it is not safe;
(c) ensure that any outrigger is fully extended and secured when it is required to;
(d) carry out lifting operations only if he has been briefed by the lifting supervisor on the lifting plan;
(e) ascertain the weight of the load, and not lift when a signalman is required and a clear signal has not been given;
(f) operate the crane in a safe manner (e.g. no load over public area, no operation within 3 metres of a live overhead power line, no dragging or pulling of a load, and securely block a crane parked on a slope) and report any failure or malfunction of the crane to the lifting supervisor and make a record of the failure or malfunction in the crane's log book or log sheet.

The OOC Regs also identified other lifting personnel, including the lifting supervisor, rigger and signalman. The lifting supervisor is to: coordinate all lifting activities, ensure only suitable and appointed personnel are involved in the lifting operations, ensure that the ground conditions are safe, brief all lifting personnel on the lifting plan and take measures to rectify all unsatisfactory or unsafe conditions to ensure that the lifting operation can be conducted safely.

Only approved crane contractors (ACC) are allowed to install, repair, alter or dismantle mobile cranes or tower cranes. ACCs need to be competent in their duties and only companies approved by the Commissioner for WSH can become an ACC. The ACCs need to be familiar with the crane manufacturers' instructions and manuals and they must follow the instructions and manuals strictly. The owner of the cranes must ensure that every crane is tested and certified by an AE.

9.9 Workplace Safety and Health (Safety and Health Management System and Auditing) Regulations 2009

The Workplace Safety and Health (Safety and Health Management System and Auditing) Regulations 2009 (SHMS Reg) provide further details on the requirements for WSH Management Systems (WSHMS) and auditing on WSHMS. In addition, Part II of the SHSM Reg provided details on the registration of WSH auditors.

The Second, Third and Fourth Schedules of the SHMS Reg are summarised in Table 9.2. These are high risk factories that are required to put in place formal WSHMS and have periodic audits. In addition, construction sites with contract sum of S$30 million or more and shipyards with 200 or more employees are required to conduct an internal review of their WSHMS every 6 months and 12 months, respectively. The frequencies of audits and internal reviews are generally based on the past WSH performance and risk posed by the different industries. The occupier is required to engage the auditor to audit the WSHMS.

Table 9.2   Summary of audit requirements for different types of workplaces in Singapore

Type of Workplace	Definition	Condition to Require WSHMS	Condition to Require Audit by Approved Auditor	Frequency of Audit by Approved Auditor	Frequency of Internal Review if Audit is Required
1. Construction worksite	Any premises where any building operation or works of engineering construction is or are being carried out by way of trade or for purposes of gain, whether or not by or on behalf of the Government or a statutory body, and includes any line or siding (not forming part of a railway) which is used in connection with the building operation or works of engineering construction.	All worksites	Contract sum of $30 million or more	At least once every 6 months	At least once every 6 months

(Continued)

Table 9.2   (*Continued*)

Type of workplace	Definition	Condition to require WSHMS	Condition to Require Audit by Approved Auditor	Frequency of Audit by Approved Auditor	Frequency of Internal Review if audit is required
2. Shipyard	Any yard (including any dock, wharf, jetty, quay and the precincts thereof) where the construction, reconstruction, repair, refitting, finishing or breaking up of ships is carried out, and includes the waters adjacent to any such yard where the construction, reconstruction, repair, refitting, finishing or breaking up of ships is carried out by or on behalf of the occupier of that yard	All shipyards	200 or more persons are employed	At least once every 12 months	At least once every 12 months

Table 9.2   (*Continued*)

3.	Metalworking factory	Any factory engaged in the manufacturing of fabricated metal products, machinery or equipment	100 or more persons are employed	100 or more persons are employed	At least once every 12 months	Not applicable
4.	Oil and gas plant	Any factory engaged in the processing or manufacturing of petroleum, petroleum products, petrochemicals or petrochemical products	All oil and gas plants	All oil and gas plants	At least once every 12 months	Not applicable
5.	Storage of hazardous substances	Any premises where the bulk storage of toxic or flammable liquid is carried on by way of trade or for the purpose of gain	All storage with capacity of 5,000 or more cubic metres for such toxic or flammable liquid	All storage with capacity of 5,000 or more cubic metres for such toxic or flammable liquid	At least once every 24 months	Not applicable

(*Continued*)

Table 9.2 (*Continued*)

Type of workplace	Definition	Condition to require WSHMS	Condition to Require Audit by Approved Auditor	Frequency of Audit by Approved Auditor	Frequency of Internal Review if audit is required
6. Chemical plant	Any factory engaged in the manufacturing of chemicals.	Any factory engaged in the manufacturing of: (a) fluorine, chlorine, hydrogen fluoride or carbon monoxide; or (b) synthetic polymers	Any factory engaged in the manufacturing of: (a) fluorine, chlorine, hydrogen fluoride or carbon monoxide; or (b) synthetic polymers	At least once every 24 months	Not applicable
7. Pharmaceutical plant	Any factory engaged in the manufacturing of pharmaceutical products or their intermediates.	All pharmaceutical plants	All pharmaceutical plants	At least once every 24 months	Not applicable
8. Semiconductor plant	Any factory engaged in the manufacturing of semiconductor wafers	All semiconductor plants	All semiconductor plants	At least once every 24 months	Not applicable

The SHMS Reg also spells out the role of a WSHMS auditor, which includes audit of any risk assessment (RA) relating to the workplace or the work carried out in that workplace, any work process at the workplace and the workplace. In an audit, the auditor's duties include:

(a) auditing the workplace in such manner as the Commissioner may determine;
(b) submitting an audit report to the occupier of the workplace upon completion of the audit indicating the findings and recommendations in the report;
(c) advising the occupier of the workplace to take immediate action to remedy any unsafe condition or unsafe work practice found during the audit that may result in imminent danger to the safety and health of persons at work; and
(d) reporting to the Commissioner the unsafe condition or unsafe work practice referred to in sub-paragraph if the occupier:

   (i) refuses to take immediate remedial action; or
   (ii) needs more than a day to remedy the unsafe condition or unsafe work practice.

The WSHMS auditor and occupier must ensure that they do not have any conflict of interest so as to ensure the credibility and independence of the audit.

9.10 Workplace Safety and Health (Work at Heights) Regulations 2013

Falling from height is one of the most common reason for workplace fatalities. Thus, the Workplace Safety and Health (Work at Heights) Regulations 2013 ("WAH Reg") is an important legislation that all workplaces must consider. The WAH Reg follows the hierarchy of control, which advocates the importance of avoidance of WAH whenever possible. The occupier of the following workplaces must have a fall prevention plan:

(a) Any worksite.
(b) Any shipyard.
(c) Any factory engaged in the processing or manufacturing of petroleum, petroleum products, petrochemicals or petrochemical products.
(d) Any premises where the bulk storage of toxic or flammable liquids is carried on by way of trade or for the purpose of gain and which has a storage capacity of 5,000 or more cubic metres for such toxic or flammable liquids.
(e) Any factory engaged in the manufacturing of:

    (a) fluorine, chlorine, hydrogen fluoride or carbon monoxide; or
    (b) synthetic polymers.

(f) Any factory engaged in the manufacturing of pharmaceutical products or their intermediates.
(g) Any factory engaged in the manufacturing of semiconductor wafers.
(h) Any factory not falling within any of the classes of workplaces described in paragraphs 1 to 7 (above), and in which 50 or more persons are employed.

The fall prevention plan is described in the Code of Practice for Working Safety at Heights (Workplace Safety and Health Council, 2013).

The responsible person (employer and/or principal) of any person conducting WAH must ensure that the person is trained to WAH. At the same time, the WAH must be under the "immediate supervision" of a competent person. This means, to this author, that the WAH supervisor should be able to intervene immediately if the workers are working unsafely.

The occupier must ensure that all open sides or openings that can cause persons to fall 2 metres or more are covered or guarded by effective guard rails or barriers to prevent falls. If the guard rails or barriers are removed to facilitate work, they must be reinstated as soon as possible, and the workers exposed to falls from height due to the removal of the guard-rails or barriers must be protected by a travel restraint (worker is restrained from reaching the edge or opening) or a fall arrest system (worker's fall can be arrested before he hits the ground or obstacle). The occupier must also ensure that the top guard rail is at least one metre tall and the vertical distance between any two adjacent guard rails is less than 600 mm wide. The guard rail or barrier must also be of good construction, sound material and adequate strength to withstand the impact during the course of fall.

The design of fall arrest and travel restraint systems can be based on Singapore Standard 607:2015 Specification for design of active fall-protection systems (SPRING Singapore, 2015). The fall arrest system must be safe for use, i.e. the force applied on the user body must be less than 6 kN and the user must not hit any obstacle or the ground during the fall. The responsible person overseeing the WAH must appoint a competent person to inspect the anchorage and anchorage line of the travel restraint and fall arrest systems. The inspection by the competent person must be conducted at the start of every shift.

The occupier must ensure that there is always a safe means of access and egress. This includes the staircases that must be effectively barricaded to prevent falls of more than two metres. The openings between the platform and a hoist (aka teagle) must also be securely fenced and provided with secure handholds at the opening or doorway.

Working on roofs, fragile surfaces, and ladders are highly dangerous forms of WAH and they are highlighted in the WAH Reg. For a fixed vertical ladder above nine metres, there must be an intermediate landing place to ensure that the continuous vertical distance does not exceed nine meters.

"Hazardous work at height" refers to work:

(a) in or on an elevated workplace from which a person could fall a distance of more than 3 metres;

(b) in the vicinity of an opening through which a person could fall a distance of more than 3 metres;

(c) in the vicinity of an edge over which a person could fall a distance of more than 3 metres;

(d) on a surface through which a person could fall a distance of more than 3 metres; or

(e) in any other place (whether above or below ground) from which a person could fall a distance of more than 3 metres.

All hazardous WAH requires a permit-to-work system to be set up and implemented by the occupier.

The WAH Reg also regulates industrial rope access systems, making requirements like having two independent anchorage lines compulsory and requiring the responsible to engage a professional engineer to design the anchorage and anchorage line of an industrial rope access system.

9.11 Workplace Safety and Health (Workplace Safety and Health Committees) Regulations 2008

The Workplace Safety and Health (WSH Committees) Regulations 2008 ("WSH Committees Regs") stipulates the formation, meetings and functions of a WSH committee, as highlighted in Section 29(1) of the WSHA. The WSH Committee is required in all factories (high-risk workplaces, as defined in the WSHA) and the occupier of the factory must set it up. The WSH committee must be chaired by a competent person appointed by the occupier. The secretary is usually the WSHO and there must be members who are representatives of employees and the management. There must always be more employee representatives in the WSH committee than employer representatives.

The WSH committee must meet at least once a month during office hours to discuss WSH matters and the meetings must be recorded in minutes. The committee must conduct inspections at least once a month. The inspection findings must then be discussed and recorded. Follow-up actions must be implemented. The WSH committee must also conduct an inspection after any accident or dangerous occurrence and the WSHO must conduct an investigation into the incident and furnish an investigation report to the WSH committee chairman. The post-incident inspection of the WSH committee must be discussed at a meeting and a report recording the WSH issues and recommendations must be produced. The occupier must then evaluate the report and take the necessary actions.

The WSH committee also has other functions, such as organising promotion activities to improve WSH, and setting WSH guidelines for the factory. For the WSH committee to be effective, the members must be trained in WSH. The WSH Committee has the following powers:

(a) to enter, inspect and examine the workplace at any reasonable time;
(b) to inspect and examine any machinery, equipment, plant, installation or article in the workplace;
(c) to require the production of workplace records, certificates, notices and documents kept or required to be kept under the Act, including any other relevant document, and to inspect and examine any of them;
(d) to make such examination and inquiry of the workplace and of any person at work at that workplace as may be necessary to execute his duties;
(e) to assess the levels of noise, illumination, heat or harmful or hazardous substances in the workplace and the exposure levels of persons at work therein;
(f) to investigate any accident, dangerous occurrence or occupational disease that occurred within the workplace.

9.12 Workplace Safety and Health (Workplace Safety and Health Officers) Regulations

A WSHO must have completed the WSQ Specialist Diploma in WSH and have at least two years of practical experience relevant to WSH. In addition, the WSHO must also pass an interview conducted by the Commissioner (MOM). The following workplaces must employ a WSHO:

(a) Shipyards in which any ship, tanker and other vessels are constructed, reconstructed, repaired, refitted, finished or broken up.
(b) Factories used for processing petroleum or petroleum products.
(c) Factories in which building operations or works of engineering construction of a contract sum of S$10 million or more are carried out.
(d) Any other factories in which 100 or more persons are employed, except those which are used for manufacturing garments.

The duties of a WSHO include:

(a) assisting the occupier of the workplace or other person in charge of the workplace to identify and assess any foreseeable risk arising from the workplace or work processes therein;
(b) recommending to the occupier of the workplace or other person in charge of the workplace reasonably practicable measures to eliminate any foreseeable risk to any person who is at work in that workplace or may be affected by the occupier's undertaking in the workplace;
(c) where it is not reasonably practicable to eliminate the risk referred to in sub-paragraph (b), recommending to the occupier of the workplace or other person in charge of the workplace:

   i.   such reasonably practicable measures to minimise the risk; and

   ii.   such safe work procedures to control the risk; and

(d)   assisting the occupier of the workplace or other person in charge of the workplace to implement the measure or safe work procedure referred to in sub-paragraph (b) or (c), as the case may be.

As can be observed, the WSHO is predominantly meant to assist and advice the occupier. To perform their duties, the WSHO is given the same set of powers as the WSH committee.

References

SPRING Singapore. (2015). *SS 607: 2015 Specification for design of active fall-protection systems*. Singapore: SPRING Singapore.

Tranter, M. (2004). *Occupational Hygiene and Risk Management — Megan Tranter — 9781741143294 — Allen & Unwin — Australia* (2nd ed.). Sidney: Allen & Unwin. Retrieved from https://www.allenandunwin.com/browse/books/general-books/business-management/Occupational-Hygiene-and-Risk-Management-Megan-Tranter-9781741143294.

Workplace Safety and Health Council. (2013). *Code of Practice for Working Safely at Heights* (Second Rev). Singapore.

Workplace Safety and Health Council. (2014). *Code of Practice on Safe Lifting Operations in the Workplaces*. https://www.wshc.sg/files/wshc/upload/cms/file/2014/Code_of_Practice_Safe_Lift-ing_Operations_Revised_2014.pdf. (accessed March 27, 2018).

Workplace Safety and Health Council. (2017). Asbestos. https://www.wshc.sg/wps/portal/!ut/p/a1/jY89D4lwEIZ_iwNr7_gQjVvjIFG-MA6jQxYCpB-VMoKRX-vsiqqLfd5XnyvgcMEmB11pUiM6WqM_namX8JD57n0Ah3my-C2kTp7DOKFa0e-NwDpNLA-uf_5ODEUf_InYN-QscElfInYAhNS5eO7Ka-1zdymAaX7jmmvy0MO5M-KZpVxZa2Pc9EUoJyclVVaQVFn6yCtUaSN5gaKp-jgve57EI6ewL. (accessed March 19, 2018).

CHAPTER 10

Accident Case Studies

10.1 Introduction

In this chapter, three accident case studies are presented to facilitate discussions about key Workplace Safety and Health (WSH) management concepts presented in the earlier chapters. At the same time, relevant WSH legislations will also be highlighted.

10.2 Case A: Crane Collapse

10.2.1 Background

This case is taken from *Jurong Primewide Pte Ltd v Moh Seng Cranes Pte Ltd and others [2014] 2 SLR 360*. With reference to Figure 10.1, Jurong Primewide Pte Ltd ("JPW") was the main contractor to build a seven-storey multi-user business park development for Crescendas Bionix Pte Ltd ("Crescendas"). JPW had a crane rental agreement with Hup Hin Transport Co Pte Ltd ("Hup Hin") on a per call basis and Hup Hin had a hiring contract with Moh Seng Cranes Pte. Ltd. ("Moh Seng") to hire Moh Seng's mobile cranes whenever required. Moh Seng was the owner of the damaged crane and the employer of the crane operator. MA Builders Pte Ltd ("MA") was JPW's subcontractor responsible for structural, architectural and external works.

10.2.2 Accident detail

On 10 June 2010, MA requested JPW to provide a mobile crane to lift some rebars. Thus, JPW contacted Hup Hin to deliver a

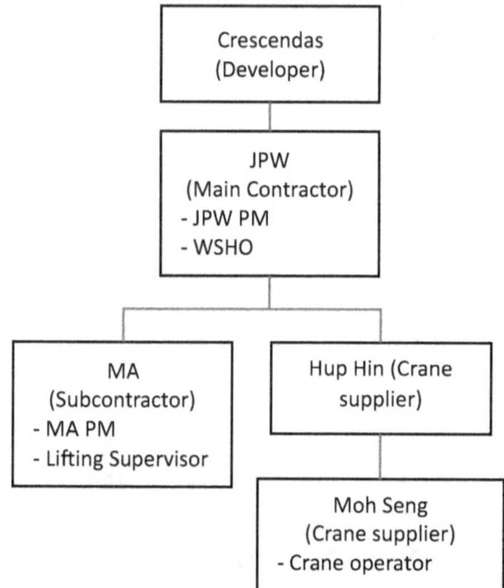

Figure 10.1   Relationship between different parties of Case A

50-tonne mobile crane to the worksite the following day. As Hup Hin did not have a 50-tonne mobile crane on 11 June 2010, they contacted Moh Seng for the crane.

On 11 June 2010 the crane operator employed by Moh Seng, arrived at the site with the 50-tonne mobile crane. He reported to the lifting supervisor employed by MA, who instructed the crane operator to park at the washing bay area near the guardhouse, which is in the vicinity of a manhole covered by soil. The crane operator expressed concerns that the designated location might not be able to take the weight of the crane safely. The lifting supervisor assured the crane operator that the location had "hard flooring" and could support the crane's weight. The crane operator was not convinced, and he expressed his concerns to JPW's workplace safety and health officer (WSHO). The WSHO discussed with the lifting supervisor and then reassured the crane operator that the ground was able to take the loading imposed by the crane. Thus, the crane operator proceeded to deploy the crane in accordance to the lifting supervisor's instructions.

When the crane operator swung the boom from the left front of the crane towards the left back outrigger, the left back outrigger broke through the cover of a concealed manhole causing the crane to collapse. The manhole had been covered by layers of brown soil and was not visible just prior the accident.

It was established that JPW knew about the manhole by 12 August 2008, but did not take reasonable care to properly mark and cordon off the manhole. They also did not direct the sub-contractors to take the appropriate action. JPW indicated that the manhole was usually visible and so "obvious to any person or passer-by" that there would not have been "the need of [sic.] any barricade, warning sign or warning". They also blamed MA for allowing soil to accumulate over the manhole due to their excavation work. However, the Judges found that the sheer knowledge of the presence of the manhole would have made it foreseeable that the accident could happen on a busy site with heavy equipment and machinery. Thus, JPW's negligence was *inexcusable*.

Nevertheless, MA, as subcontractor, was also expected to ensure that the lifting operation was conducted safely. This is especially the case because MA was the employer of the lifting supervisor who was instructing the lifting work. Furthermore, MA was fully informed of the location of the manhole because they conducted investigation and piping works in the manhole prior to the accident. MA's project manager (PM) had also warned the lifting supervisor about the manhole, and the lifting supervisor acknowledged that he was aware of the presence of the manhole. Despite this knowledge, MA's lifting supervisor commenced lifting operations without a valid permit-to-work (PTW).

10.2.3 Event Causation Technique analysis

Figures 10.2 and 10.3 show the Why-analysis and Event Causation Technique (ECT) diagram for the accident, respectively. As seen

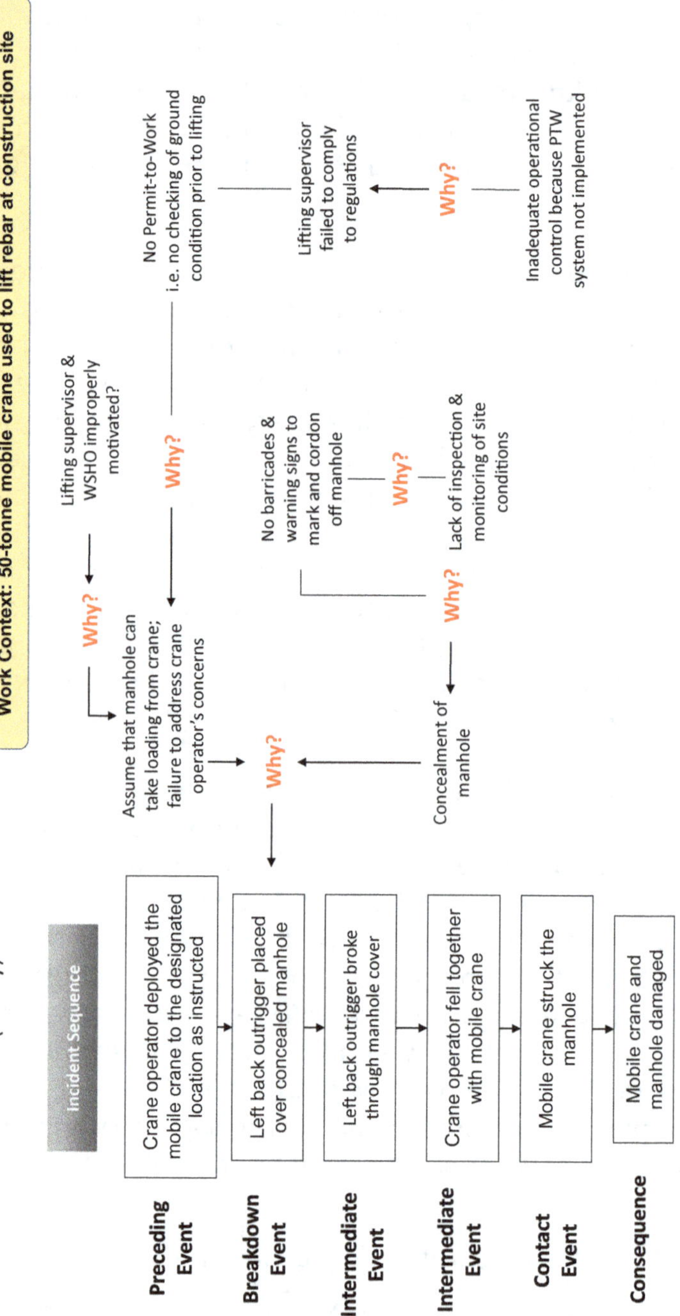

Figure 10.2   Why-analysis for Case A

in the figures, the breakdown event is, "Left back outrigger placed over concealed manhole". The direct cause of the accident was the decision made by the lifting supervisor and JPW WSHO to locate the crane near the washing bay. This decision was probably based on their assumption that the manhole cover could take the load from the outrigger and/or the outrigger was not placed directly over the concealed manhole. The lifting supervisor and WSHO could have been improperly motivated, but there was a lack of information on the reason why the two safety personnel made the decision despite concerns from the crane operator. The concealment of the manhole was also a direct cause leading to the outrigger being placed over the concealed manhole.

One control that could have been implemented to prevent the breakdown event is a PTW system to ensure that legal requirements such as checking on ground conditions were conducted prior to lifting. Reg 12 of the Workplace Safety and Health (Construction) Regulations 2007 (No S 663/2007) required a PTW to be first issued by the PM of the worksite before high-risk construction work could be conducted. At the time of the accident, Reg 20(3)(c) of the Factories (Operations of Cranes) Regulations 1998 ("the Regulations") was still in force and it stated that a lifting supervisor has to "ensure that the ground conditions are safe for any lifting operation to be performed by any mobile crane". This was also highlighted in the Singapore Standard SS 536 2008 Code of Practice for the safe use of mobile cranes, which specifically identified manholes as one of the possible "underground hazards" that must be checked for.

Another possible control that could have prevented the outrigger from being placed near or over the manhole is to clearly mark and cordon off the manhole area. This could also potentially prevent the manhole from becoming concealed by the soil during excavation.

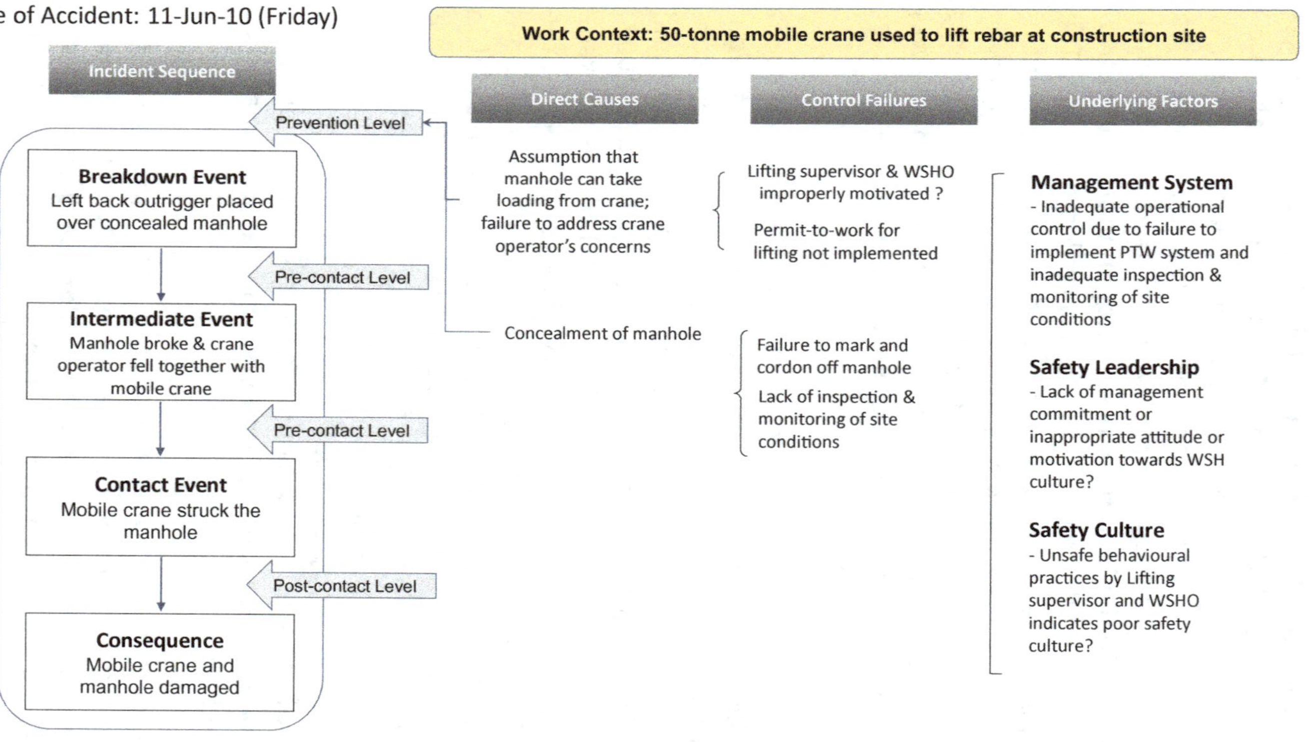

Figure 10.3   ECT diagram for Case A

Through the actions of the lifting supervisor and WSHO, who ignored the concerns of the crane operator about the ground condition, it is inferred that the safety culture and leadership of the site was probably lacking prior to the accident. In addition, with the failure to implement the PTW system for high-risk activities like lifting, and the lack of site inspection and monitoring, it can be inferred that the level of implementation of the operations control element of the management system was probably inadequate.

10.2.4 Discussion

This discussion will focus on the legislative implications of the case. WSH management system concepts will be discussed at the end of the chapter.

In their judgement, the Appeal Judges highlighted the intention of the Workplace Safety and Health Act (WSHA), which is to create a system of accountability by defining the duties of different stakeholders. As then Minister of Manpower Dr Ng Eng Hen, during the second reading of the Workplace Safety and Health Bill 2005 (Bill 36 of 2005) (Singapore Parliamentary Debates, Official Report (17 January 2006) vol 80) at cols 2208–2210, explained:

> "...this Bill will better *define persons who are accountable, their responsibilities and institute penalties* which reflect the true economic and social cost of risks and accidents. Penalties should be sufficient to deter risk-taking behaviour and ensure that companies are *proactive in preventing incidents.* Appropriately, companies and persons that show *poor safety management* should be penalised even if no accident has occurred.
>
> This Bill will put into place a new and more effective framework to reduce accidents at the workplace — to bring about a quantum improvement in OSH (occupational safety and health) standards and to achieve our intermediate goal of halving the present occupational fatality rate by 2015."

Accordingly, contractors and subcontractors, i.e. JPW and MA respectively, are two of the entities which the WSHA seeks to "increase direct liability on for workplace safety". Main contractors and subcontractors have direct "operational control" of workplaces and they have tremendous responsibilities to ensure a safe working environment at construction sites. Even though the main contractor, as the Occupier of the worksite, has a duty under Section 11 of the Workplace Safety and Health Act (Chapter 354A) to ensure, so far as is reasonably practicable, that the workplace, all means of access to or egress, and any machinery, equipment, plant, article or substance are safe, principals (MA in this case) and employers have an important role because they direct the manner of work. Dr Ng Eng Hen, in the second reading of the Workplace Safety and Health Bill 2005, also explained how the WSHA sought to impose duties on principals engaging contractors for specialised tasks (at col 2209):

> "Traditionally, a principal who engages a contractor would be engaging the specialist services of the contractor, and would not be directing the contractor on how to do the work. However, today the situation is different. Principals often engage "contractors" and third-party labour not for their specialist expertise, but *precisely so that they can avoid entering into a direct employment relationship, for organisational or other reasons.* In such situations, the principal, in terms of supervision, takes on the role of an employer. *The Bill thus places on him responsibility for the worker's safety and health as if he were the employer.* If this were not the case, then the *duties under the Act could be simply circumvented by a careful crafting of the legal relationship.*"

Thus, JPW, as the main contractor, had to ensure that "hazards such as manholes were identified properly through site surveys, and then to under-take ground improvements to ensure that these underground hazards did not continue to remain hazardous". Even though MA was "not contributorily negligent in causing some soil run-off to cover the manhole due to its nearby excavation works", it is still responsible for the accident because the "lifting operation that caused the accident was within MA's scope of work" (the lifting supervisor, riggers and signalmen were employed by MA), "MA knew about the manhole as well", and "MA was contractually responsible and deemed fully informed of the conditions at the worksite".

The lifting supervisor is an important safety personnel during a lifting operation, but in this case the lifting supervisor failed to perform his duties specified in Reg 20(3)(c) of the Factories (Operations of Cranes) Regulations 1998, which states that a lifting supervisor has to "ensure that the ground conditions are safe for any lifting operation to be performed by any mobile crane". Furthermore, the Singapore Standard SS 536 2008 Code of Practice for the safe use of mobile cranes, required the lifting supervisor to (italicised text are relevant duties for the case study):

(a) *co-ordinate and be present to supervise all lifting activities and ensure that the lifting operation is carried out safely;*
(b) ensure that only registered crane operators, appointed riggers and appointed signalmen participate in any lifting operation involving the use of a mobile crane;
(c) *ensure that the ground conditions are safe for any lifting operation to be performed by any mobile crane;*
(d) ensure that there is a set of safe lifting procedures for any lifting operation of a mobile crane;
(e) brief all crane operators, riggers and signalmen on the safe lifting procedures referred to in (d);
(f) *take measures to rectify the unsatisfactory or unsafe conditions that are reported by any crane operator or rigger.*

The Judges were especially critical on the lifting supervisor because though he was aware that there was a concealed manhole, he failed to ensure that the hazard was effectively managed. Worse still, the lifting supervisor failed to take any measures to address the crane operator's concerns about the safety of the ground conditions. Thus, the Judges "had no hesitation in finding that MA was liable in negligence for the damage caused during the accident".

Finally, the Judges made the following decision: "both JPW and MA were liable in negligence to Moh Seng, and that MA did breach the subcontracts that it had entered into with JPW, we felt that an apportionment of 60% to JPW and 40% to MA would be just in all the circumstances. As the main contractor, as well as the occupier under the WSHA, JPW had to bear the bulk of the responsibility for failing to identify an underground hazard like the manhole and for not taking measures to ensure that it ceased to be an unknown danger. However, we were also cognisant of the fact that MA was largely in charge of the works that were in the area, and specifically, the operation of lifting steel rebars that [the crane operator's] crane was engaged in. Also, the Lifting Supervisor was under MA's employment and had specific duties under the WSH Regime which he failed to fulfil."

10.3 Case B: Falling from Loading Platform During Lifting

10.3.1 Background

This case is taken from *Public Prosecutor v GS Engineering & Construction Corp [2016] SGHC 276.* With reference to Figure 10.4, GS Engineering & Construction Corp (GSECC) was the main contractor to construct two towers (Tower A and Tower B) for Jurong Town Corporation. The two towers were to be 11 and 18 storeys high respectively. GSECC subcontracted the structural works (including formwork) for Tower A to Zhang Hui Construction Pte Ltd ("Zhang Hui"). The project started on 23 November 2011 and was meant to be completed by 23 March 2014.

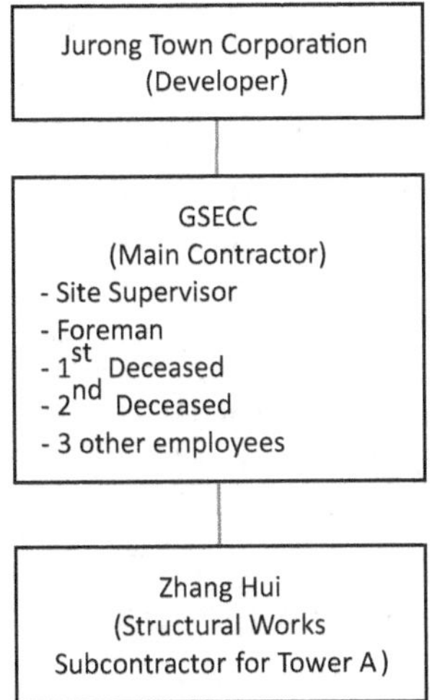

Figure 10.4   Parties involved in Case B

10.3.2 Accident detail

On 22 January 2014, GSECC employees were initially planning to shift the loading platform from the tenth floor of Tower B to the eighth floor of Tower A. Loading platforms are typically positioned on the edge of buildings to facilitate the lifting of materials and items to and from different levels of the building. When employees from Zhang Hui requested for the GSECC site supervisor to move the air compressor from the seventh floor of Tower A to the eighth floor of the same tower using the loading platform, the site supervisor instructed the GSECC foreman to assist Zhang Hui on this task. The site supervisor instructed the foreman "not to install the loading platform at the seventh storey of Tower A, but to simply suspend it by a tower crane." It is presumed that the site supervisor wanted to save time so that the time taken for the additional task of collecting the air compressor at level seven can be minimised.

At about 11.50am, the foreman with five workers, including the two deceased, commenced the task of shifting the loading platform from Tower B to the seventh storey of Tower A. As

Zhang Hui's workers were at lunch, the GSECC foreman proceeded with his workers. The air compressor was mounted on a steel frame with four wheels, but the rear wheels were smaller and could not roll onto the suspended loading platform. In addition, during the loading process, the suspended platform tilted. One of the deceased workers and his co-workers informed the foreman that it was unsafe to push the air compressor onto the suspended platform. However, the foreman instructed them to continue with the task.

The workers then used a galvanised pipe to pivot the air compressor onto the loading platform. After several attempts, the air compressor was pushed onto the platform, but it rolled towards the two deceased who were on the platform, causing the platform to tilt. The two deceased were not able to evade the air compressor in time and fell together with it.

The air compressor landed on another loading platform two levels below. The two deceased fell to the ground level and were pronounced dead by the paramedics attending to the accident.

10.3.3 Event Causation Technique analysis

The incident sequence is summarised in the why-analysis (Figure 10.5) and ECT diagram (Figure 10.6). The breakdown event is "Platform tilted & air compressor rolled towards 2 deceased" and the contact event is "2 deceased hit the ground".

The breakdown event happened because prior to it, the workers were pushing the air compressor on wheels onto the platform and the loading platform was suspended by the tower crane using four lifting chain slings. Furthermore, the foreman failed to address the concerns of the workers who felt that it was unsafe to proceed with the work because the platform tilted when they were trying to load the compressor onto the

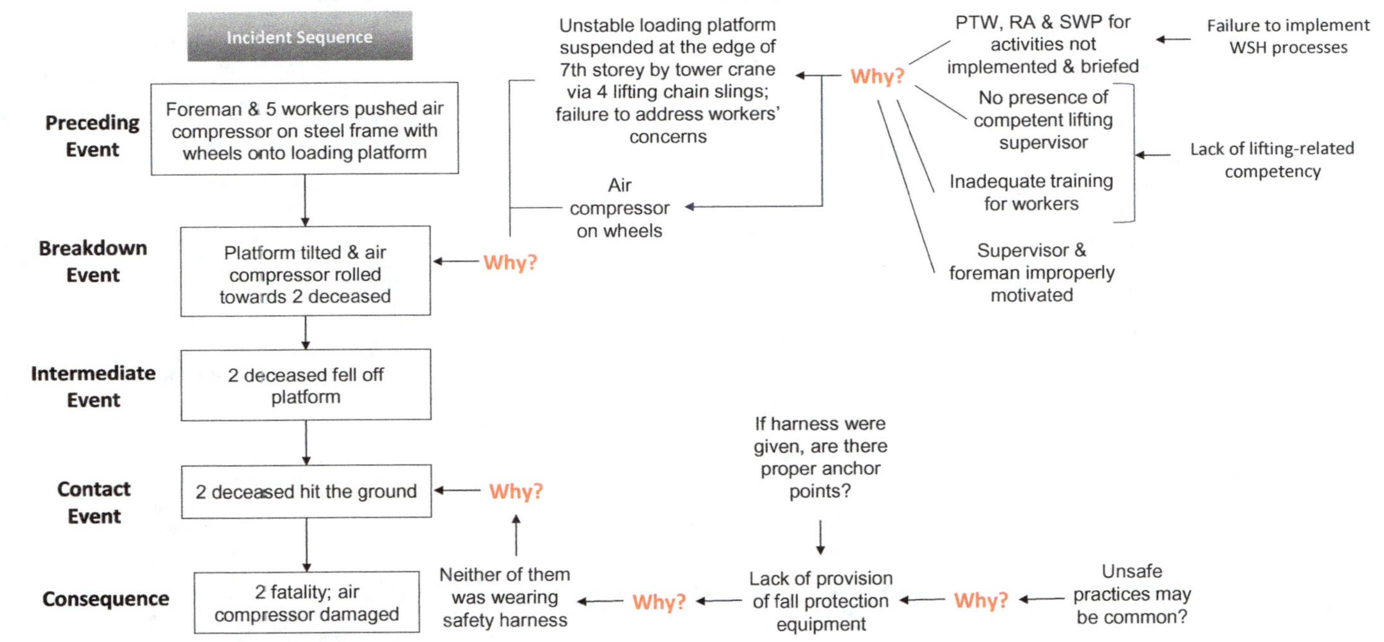

Figure 10.5   Why analysis for Case B

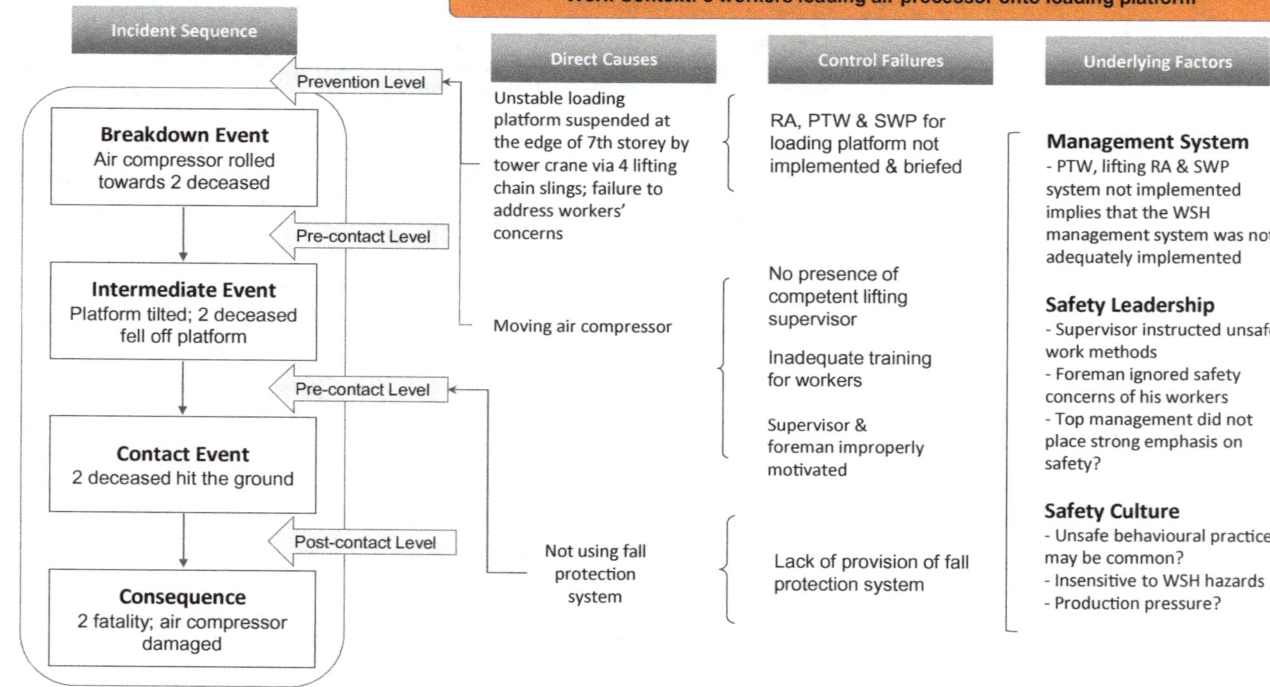

Figure 10.6   ECT diagram for Case B

platform. Effective training on the safe use of the loading platform, the presence of a competent lifting supervisor, the implementation and briefing of the PTW, risk assessment (RA) and safe work procedure (SWP) for the lifting and use of the loading platform should have resulted in the removal of the unsafe work method and hence prevented the breakdown event. Even though a foreman was present, he was likely overly focused on the completion of the task and failed to consider the safety of the workers. Similarly, the supervisor had probably instructed the unsafe method of work to save time. In addition, the lack of provision of a fall protection system (e.g. fall arrest anchor, lifeline and harness) to the workers on the day of the accident means that it is very unlikely to prevent the fatality when the breakdown event happened.

There is a lack of information on the state of the safety culture of the project but based on the actions of the foreman and supervisor it seems that frontline leaders (supervisors and foremen) may not place due emphasis on safety. Similarly, several operational controls like PTW, RA and SWP not being adequately implemented are signs that the management system is ineffective, indicating inadequacies in the organisation's safety culture and safety leadership.

10.3.4 Discussion

This discussion will focus on the legal implications of the case. WSH management implications will be discussed at the end of the chapter.

This case is a landmark case because it was the first time a WSHA case was presented to the High Court. Furthermore, the Prosecution disputed the District Judge's sentencing and proposed a new sentencing framework. One of the key rationales for the Prosecution's proposal was that the sentencing for previous cases were too low, with the majority falling below 30%

of the maximum sentence of the S$500,000 maximum fine prescribed in the WSHA. The relatively low sentencing was contrary to the legislative intent of the WSHA, which the then Minister for Manpower, Dr Ng Eng Hen, described at the second reading of the Workplace Safety and Health Bill (Bill 36 of 2005) ("the Bill") as follows:

> "Three fundamental reforms in this Bill will improve safety at the workplace. First, the Bill will strengthen proactive measures. Instead of reacting to accidents after they have occurred, which is often too little too late, we should reduce risks to prevent accidents. To achieve this, all employers will be required to conduct comprehensive risk assessments for all work processes and provide detailed plans to minimise or eliminate risks.
>
> Second, industry must take ownership of occupational safety and health standards and outcomes to effect a cultural change of respect for life and livelihoods at the workplace...
>
> Third, this Bill will better define persons who are accountable, their responsibilities and *institute penalties which reflect the true economic and social cost of risks and accidents. Penalties should be sufficient to deter risk-taking behaviour and ensure that companies are proactive in preventing accidents.* Appropriately, companies and persons that show poor safety management should be penalised even if no accident has occurred."

Furthermore, Dr Ng explained the reason for the need for higher penalties for poor safety management and performance:

> "Even as we work with industry to build up their capabilities to improve safety and health at their workplaces, *we need to ensure that the penalties for non-compliance are sufficiently high to effect a cultural change on the ground. Penalties should be set at a level that reflects the true cost of poor safety management, including the cost of disruptions and inconvenience to members of the public which workplace accidents*

will cause. The collapse of the Nicoll Highway not only resulted in the loss of four lives, but also caused millions of dollars in property damage and led to countless lost working hours and great inconvenience to the public. The maximum penalty of [S]$200,000 under the present Factories Act is therefore inadequate."

The High Court Judge largely adopted the proposed sentencing framework, which is presented in Table 10.1. However, it should be noted that this table was updated in the subsequent case *MW Group Pte Ltd v Public Prosecutor [2019] SGHC 5* (see Table 10.2), which will be discussed herein.

The Judge also provided a detailed description of the process for sentencing by a court using Table 10.1. With reference to Table 10.1, a court will determine the appropriate starting point for the sentence by considering two principal factors: (i) the culpability of the offender; and (ii) the harm that could potentially have resulted, as described below:

1. Determine the culpability of the offender based on the following non-exhaustive factors:

    a. the number of breaches or failures in the case;
    b. the nature of the breaches;

Table 10.1   Sentencing framework determined by High Court (all monetary figures shown are in SGD)

		Culpability		
		High	**Medium**	**Low**
Potential for harm	High	$300,000 to $500,000	$150,000 to $300,000	$100,000 to $150,000
	Medium	$100,000 to $150,000	$80,000 to $100,000	$60,000 to $80,000
	Low	$40,000 to $60,000	$20,000 to $40,000	Up to $20,000

Table 10.2 Updated sentencing framework (adapted from *MW Group Pte Ltd v Public Prosecutor [2019] SGHC 5*) (all monetary figures shown are in SGD)

Potential for harm	> High	Possible to extend to maximum sentence prescribed by law			
	High	Up to $180,000 CG: $107,400	$120,000–$300,000 CG: $204,200	$200,000–$360,000 CG: $282,500	Possible to extend to maximum sentence prescribed by law
	Medium	Up to $120,000 CG: $65,800	$60,000–$200,000 CG: $124,100	$110,000–$250,000 CG: $176,900	
	Low	Up to $60,000 CG: $23,500	Up to $110,000 CG: $47,000	Up to $140,000 CG: $71,300	
		Low	Medium	High	> High
		Culpability			

*CG is "Centre of Gravity" or the starting point for the penalty within the category.

    c. the seriousness of the breaches — whether they were a minor departure from the established procedure or whether they were a complete disregard of the procedures;

    d. whether the breaches were systemic or whether they were part of an isolated incident; and

    e. whether the breaches were intentional, rash or negligent.

2. The potential harm may be assessed by considering, among other things, (i) the seriousness of the harm risked; and (ii) the likelihood of that harm arising.

3. After deriving the starting point for sentencing, the court should calibrate the sentence by taking into account the aggravating factors and mitigating factors of the case.

4. Aggravating factors include the following: (i) serious actual harm (including death) resulted; (ii) the breach was a significant cause of the harm that resulted — in this regard, a significant cause need not be the sole or principal cause of the harm, and need only be a cause that has more than minimally, negligibly or trivially contributed to the outcome; (iii) the offender had cut costs at the expense of the safety of the workers; (iv) there was a deliberate concealment

of the illegal nature of the activity; (v) there was a breach of a court order; (vi) there was an obstruction of justice; (vii) the offender has a poor record with respect to workplace health and safety; (viii) there was falsification of documentation or licences; and (ix) there was a deliberate failure to obtain or comply with relevant licences in order to avoid scrutiny by the authorities.

5. Mitigating factors may include the following: (i) the offender has voluntarily taken steps to remedy the problem; (ii) the offender provided a high level of cooperation with the authorities for the investigations, beyond that which is normally expected; (iii) there is self-reporting, cooperation and acceptance of responsibility; (iv) there is a timely plea of guilt; (v) the offender has a good health and safety record; and (vi) the offender has effective health and safety procedures in place.

For the case of *GSECC v PP*, the Judge opined that GSECC's (Respondent) culpability falls into the Medium to High category because the occupier is overall in-charge of the worksite. Concurrently, GSECC was also the employer of the two deceased and the other workers involved in the task that resulted in the accident. Furthermore, the occupier failed to perform numerous control measures that are expected of their own lifting operations and it also failed to ensure that its workers were adequately trained and had adequate fall protection. Thus, even if Zhang Hui had not asked GSECC to shift the air compressor, the occupier would have committed numerous breaches and failures. Moreover, the occupier oversees the PTW system and would have to evaluate and approve Zhang Hui's application to conduct the lifting operation.

During the trial, the Respondent attempted to shift the responsibility to its workers, but the Judge noted then Minister of Manpower, Dr Ng Eng Hen's, observations at the second reading of the Workplace Safety and Health Bill (*Singapore*

Parliamentary Debates, Official Report (17 January 2006) vol 80 at col 2205):

> "The reality is that on a day-to-day basis, safety may be the last thing on the minds of management and workers on the ground. There are deadlines to meet, monotony, apathy or lethargy to overcome, a lack of professionalism and training, unclear lines or no lines of accountability, and poor management..."

Even though workers have duties under the WSHA to cooperate with their employers, the intent of the WSHA is for the employer, occupier or other responsible persons to "ensure that its workers are trained and are mindful of their safety at the workplace, and that proper systems are in place to ensure that steps are taken to minimise risks."

In terms of potential harm, the Judge selected the "high" category because the high-risk work was conducted without a safe system of work, and because fall arrest equipment were not provided. Furthermore, many of the workers were not trained for the task. Thus, all the workers involved could have been killed. In addition, the landing platform could have failed and landed on other workers.

Therefore, based on the level of culpability and the potential for harm, the starting sentence was a fine of S$300,000.

The Judge then considered the aggravating and mitigating factors to calibrate the sentence. Aggravating factors included two lives being lost as a result of the breaches. Mitigating factors included the fact that the Respondent had pleaded guilty, cooperated with the authorities, had a good safety record, won past awards from the Land Transport Authority and MOM (two WSH awards for the accident site), and had been proactive in investigating the accident and implementing remedial actions. Finally, after weighing all the relevant factors, the Judge arrived at an appropriate sentence of S$250,000.

As indicated earlier, the judge in the subsequent case *MW Group Pte Ltd v Public Prosecutor [2019]* updated the sentencing framework (see Table 10.2). The main concern he had was the sentencing gaps between some of the cells in Table 10.1. For example, the sentencing range had a significant increase of S$40,000 when moving from "low culpability-low potential for harm" to "low culpability-medium potential for harm". To solve the problem the judge took a mathematical approach and drew a series of curves to ensure consistency in the sentencing range for each of the cells in the table. The judge simplified the mathematical graphs into the numbers in Table 10.2. Even though the values are updated, the guidance for aggravating and mitigating factors are largely similar.

10.4 Case C: Falling from Mezzanine Floor During Welding Work

10.4.1 Background

This case is based on a consultancy project that the author conducted in 2018. The client had agreed to allow the author to discuss the accident, but it must be clarified that this case study is based on the author's opinions and interpretations. Fictitious names were used to prevent any possible complications.

Figure 10.7 shows the relationship between the key parties involved in this accident case study. The developer is a public agency and they selected the main contractor to construct a two-storey multi-user motor workshop. The main contractor sub-contracted fabrication, supply, and installation of metal works to the sub-contractor. The deceased and his two co-workers were employed by a labour supplier who supplies short term workers to different construction worksites and other workplaces. In this case, the deceased and his co-workers were working for the metal work sub-contractor. The deceased was a 45-year-old male Chinese worker from China.

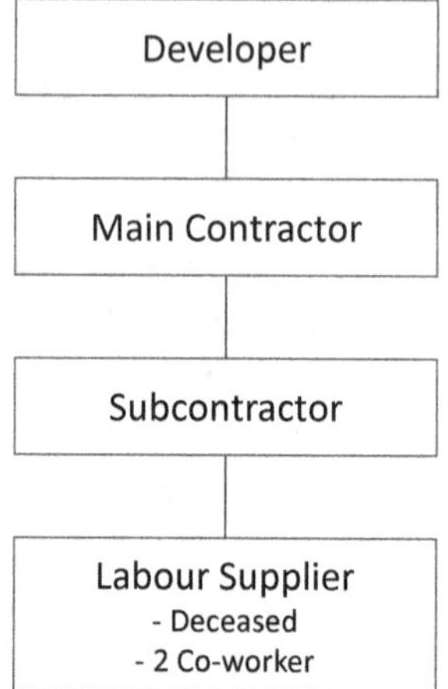

Figure 10.7   Parties involved in Case C

This case study is focused on the duties of the subcontractor, who is deemed as the principal controlling and directing the work of the deceased and his co-workers and hence had the same duties as an employer under the WSHA. The project commenced on 2 April 2012 and the sub-contractor's contract with the main contractor was made on 25 December 2012. The subcontractor was engaged to "supply and install the galvanized steel hollow section with steel plates and bolts to the mezzanine office dry wall partitions at the 1st and 2nd storeys". A mezzanine floor is an intermediate floor in a unit of a building (e.g. an apartment or a motor workshop) that does not cover the whole floor space of the unit.

10.4.2 Accident detail

On 27 April 2014, the deceased and his two co-workers were installing steel hollow sections (SHS) at the mezzanine floor located within a second level motor workshop unit under construction (see Figure 10.8). The SHS were used to construct the metal frame for the dry wall for the mezzanine floor. Each

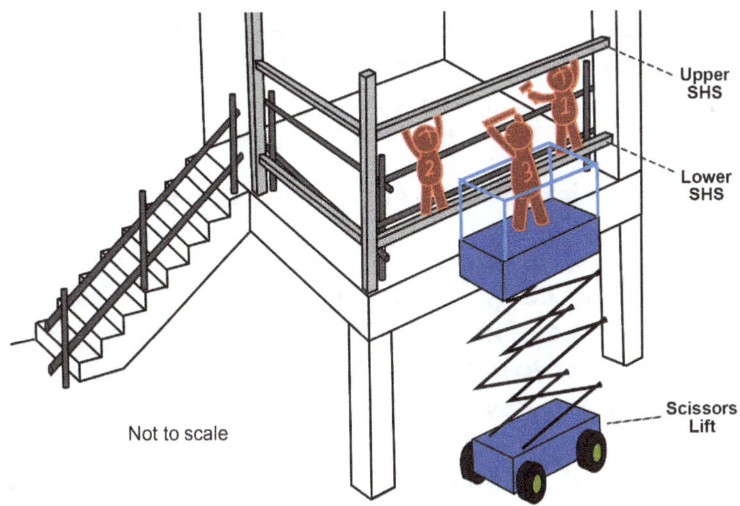

Figure 10.8   Overview of work prior to accident

metal frame for a dry wall was mainly made up of vertical columns and two horizontal beams (upper and lower SHS). The mezzanine floor was 2.8m above the floor of the unit. For the deceased, it was his first day on this site.

Prior to the accident, the deceased (worker 1 in Figure 10.8) was hammering the upper SHS (about 5.1m long) to level it while standing on the lower SHS and/or the edge barricades with his co-worker (worker 2). Even though there were some discrepancy on the position of the deceased and his co-worker, based on the details of the case, the author opined that it was most likely that they were indeed standing on the lower SHS and/or the barricades. The other co-worker (worker 3) was working from a scissors lift (a type of mobile elevated work platform (MEWP)) placed next to the mezzanine floor. The lower SHS was tack welded into place. It is noted that tack welds are meant for preliminary positioning and alignment purposes and they are not meant to take significant load. Furthermore, the barricades were not properly secured and were simply placed near the edge.

At about 3.45pm, the deceased fell from the mezzanine floor and was conveyed to the hospital where he succumbed to his injuries on 2 May 2014 at 3.45am.

10.4.3 Event Causation Technique analysis

The incident sequence is summarised in the why-analysis (Figure 10.9) and ECT diagram (Figure 10.10). The breakdown event could not be definitively confirmed, but the likely causes were "tack welded lower SHS failed" and "unsecured barricade shifted". When the lower SHS and/or the barricade moved, the deceased lost balance and fell onto the floor. The two main incident events that should be evaluated are the breakdown event and contact event.

The deceased was standing on an unsafe work platform (the tack welded lower SHS and/or the unsecured barricades) prior to the accident. It must be noted that the workers were not provided with the necessary safe working platform. According to a set of work procedure that the sub-contractor produced, the deceased and his co-workers were supposed to work from two scissors lifts, but they were only provided one scissors lift on the day of the accident, which was not enough for them to do their work properly. Furthermore, the deceased and his co-workers were not trained to operate the scissors lift that they were using. The deceased was also not trained for welding and working-at-height (WAH).

If the workers were expected to use the mezzanine floor to conduct the work, the barricades were not properly secured and positioned. Furthermore, the workers were not provided suitable step ladder or platform to reach the height of the upper SHS. Thus, the workers had to work unsafely so as to position themselves to conduct the welding work.

The sub-contractor felt that it was the workers' poor safety attitude that led to the accident. In fact, the deceased was

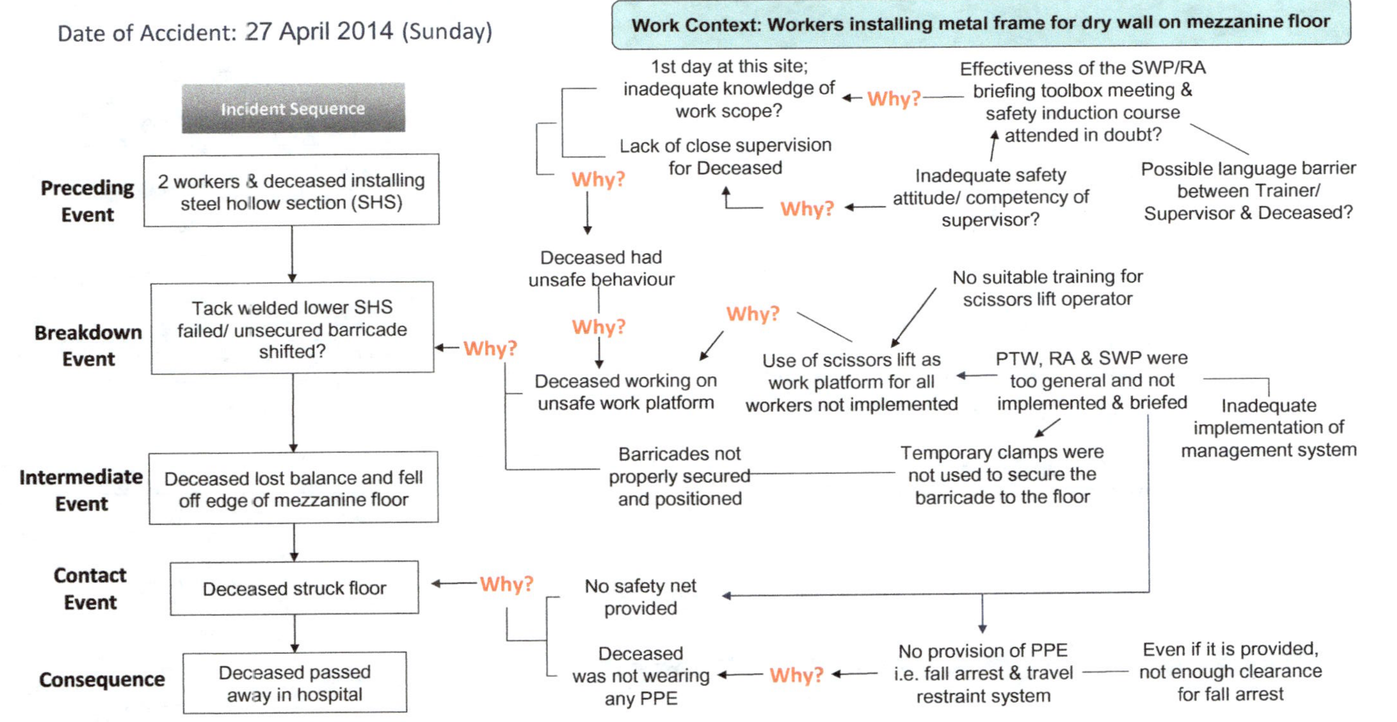

Figure 10.9   Why-analysis for Case C

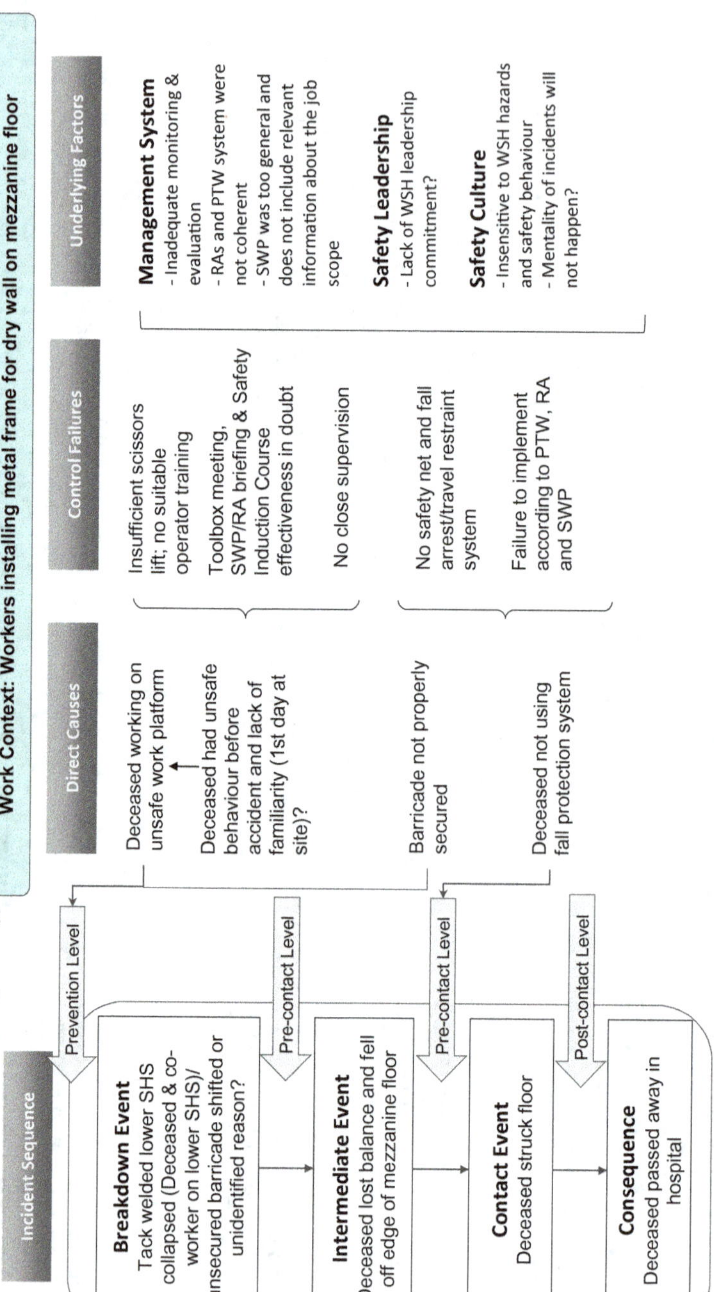

Figure 10.10   ECT diagram for Case C.

issued with an "EHS Administrative Charge" (essentially a site imposed fine) on the day of the accident. He was found by the main contractor to be working unsafely at height without proper personal protective equipment (PPE). The main contractor instructed the sub-contractor to stop the work-at-heights. The sub-contractor supervisor received the "EHS Administrative Charge".

The unsafe behaviour of the deceased could be related to the fact that it was his first day working at the site and he might be unfamiliar with the site requirements and work conditions. His safety attitude and behaviour could have been managed through briefings and trainings, and close supervision. However, these control measures were ineffective.

The site had a series of briefings and training for the deceased and his co-workers when they arrived at about 8am on the day of the accident. These included, first day orientation, a safety induction course, SWP and RA briefing and toolbox meeting. The deceased also signed on two PTW forms. All these activities happened between 8am and 9am. It was also noted that the supervisor was from India, while the deceased and co-workers were from China. Based on the author's experience, the language barrier between workers of different nationalities was (and continues to be) a major challenge in the Singapore construction industry. Thus, in view of the language barrier and limited time, it was very unlikely that the training and briefings were in-depth and effective. However, it was not established how the language barrier could have contributed to the accident.

Furthermore, knowing that the deceased was working on the site for the first time, the sub-contractor should have provided close supervision on his work. Close supervision means that the supervisor must be able to intervene in a timely fashion when the deceased was conducting his work. Close supervision was even more important after the deceased received the "admin charge" but was allowed to continue to work on high risk tasks such as work-at-height and welding. However, the supervisor did not provide close supervision on the deceased and his co-workers. If close supervision was not possible, the alternative was to disallow the deceased to continue with the high risk work. The lack of actions by the supervisor cast doubts over his safety attitude and competency.

In terms of the contact event, where the deceased struck the floor after falling off the mezzanine floor, a relevant control would have been the fall protection system. A well-designed travel restraint system, safety net or fall arrest system can possibly prevent the fatality. Figure 10.11 shows two possible alternatives for implementing a travel restraint system, which the subcontractor, together with the main contractor, could have been provided with prior to the accident. Option 1 in Figure 10.11 is an adjustable travel restraint system and option 2 is a travel restraint system of fixed length. If a fall arrest system was used, the lack of fall clearance would probably mean that a self-retracting lanyard had to be used. Even though the workers were provided with full body harnesses and lanyards, they did

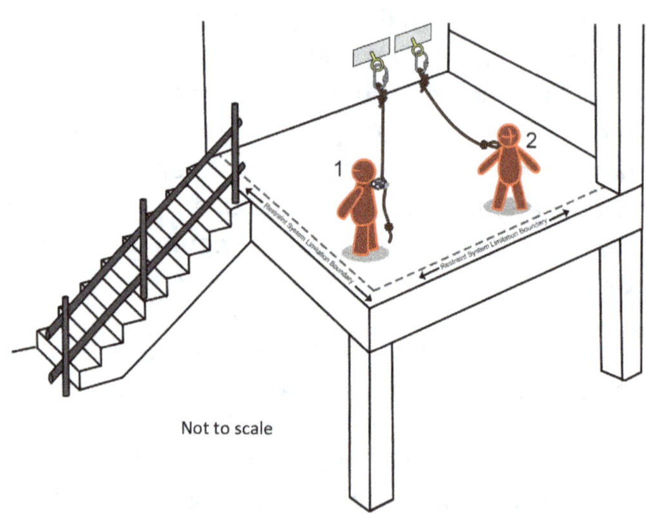

Figure 10.11 Possible travel restraint systems to protect the workers

not have the proper anchors to protect themselves and the lanyards were not suitable because the extension of the personal energy absorber during a fall arrest would mean that the clearance of 2.8m was inadequate.

The sub-contractor had a suite of documents, albeit at times incoherent, which contained many of the control measures discussed above. For example, the PTW indicated that a fall arrest system, including a lifeline, should be provided, but no fall arrest anchors or lifelines were identified or installed. It was also noted that the project manager had signed off on the PTW, but the WSHO did not sign off even though it was required in the form. The PTW also indicated requirements for a fire extinguisher, flashback arrestor, cylinder, hoses, 'O' clips, and leakage checks to be provided, but no fire extinguisher could be observed and the workers were using an electrical welding set; i.e. the flashback arrestor, cylinder, hoses, 'O' clips, and leakage checks were not relevant.

The RAs and SWPs were also in a mess. Many of the safety measures were general and not specific to the work and were not applicable. Linkages between the documents were not clearly established. In fact, they appeared to be independent of each other and were of inadequate quality. The RAs were very general and not specific to activities that the workers were conducting. They also had mistakes in terms of their calculations of risk levels.

The wide range of control failures and inadequacies in the RA, PTW system and SWPs showed that the WSH management system was inadequately implemented. Together with the apparent insensitivity to hazards and the unsafe behaviour of the workers and supervisor, it is justified to assess that safety culture was very poor. Although no information on top management's level of commitment to WSH was provided, based on the ineffective WSH management system and poor safety culture, it is hard to imagine that the top management of the sub-contractor is dedicated to WSH.

10.4.4 Discussion

The case demonstrated a severe disregard for safety by the sub-contractor. The actual manner of work of the deceased and his co-worker was very dangerous because they were not on any safe work platform and were not protected by any fall protection system or PPE. The work procedure was not in accordance to applicable safety codes, standards and guidelines referred to in the evaluation.

The deceased had displayed unsafe behaviour prior to the accident, was not trained for the welding work nor for WAH, and was not familiar with the worksite, yet he was conducting two high risk tasks concurrently. Thus, close supervision was expected, but was not implemented. The supervisor might not have been able to communicate clearly with the deceased. The unsafe behaviour that the deceased displayed may be induced by the lack of proper safety and work provisions (e.g. a raised platform for him to reach the upper SHS). The situation in this case could have induced a "necessary violation" (Reason, 1997, p. 73), where

> "non-compliance is seen as essential in order to get the job done. Necessary violations are commonly provoked by organizational failings with regard to the site, tools or equipment… In addition, they can also provide an easier way of working. The combined effect of these two factors often leads to these violations becoming routine rather than exceptional."

The deceased and his co-workers were expected to complete their work without the second scissors lift and safe work platform. In such a situation, they may have thought that it was necessary to violate WSH rules to get their work done.

The effectiveness of the toolbox meeting, briefings, first day orientation and safety induction course were doubtful because it was not possible for so many briefings and meetings to be conducted within one hour. It was also noted that the deceased was not trained for welding and WAH. His co-workers were also not trained to operate the scissors lift.

The range of WSH documents (RA, SWP, method statement and PTW) were not specific to the tasks that the deceased and his co-workers were conducting and the location that he and his co-workers were in. The controls identified were too general and there were numerous mistakes in the documents. Many of the controls identified were also not implemented. As a whole, the WSH-related documents were not coherent and were of low quality. The WSH management system was not effective as it was not conducted with due care. The documents reflected a low level of emphasis on WSH and poor safety leadership.

The sub-contractor seemed to be focused on having docu-ments to satisfy regulatory requirements, but had no intention to satisfy the spirit of the WSHA. Such is what the author would call a "paper exercise", or surface compliance, which was dis-cussed in earlier chapters. These WSH documents were meant to facilitate WSH processes. Writing the information down requires the site personnel to consider hazards, risks and con-trols more deliberately and in more detail. However, in practice, these WSH processes are frequently seen as additional tasks, and some companies look for "shortcuts" to complete them.

10.5 Discussion on Workplace Safety and Health Management Lessons

The three accident cases presented important WSH manage-ment lessons. These cases highlight the importance of frontline leadership by supervisors, foremen and site personnel. These supervisory personnel make many decisions on the ground that can save or kill workers, and their decisions will be based on their perception of what is important and what they believe is the right way to conduct their work. Applying the concepts dis-cussed in earlier chapters, to ensure that supervisory personnel adhere to WSH procedures, it is of utmost importance that managers impress upon them the importance of WSH. To do so managers must have dedicated WSH communication sessions with workers and supervisors.

At the same time managers must "give meaning" to the administrative controls such as the PTW system, RAs and SWPs, which, as can be observed in the three cases discussed in this chapter, can be easily defeated and violated. As discussed in Case C it is a common problem in workplaces, that paperwork is treated as just paperwork, i.e. the controls identified are not implemented and communicated to the workers. It is only when managers review and discuss the documents with site personnel will they see the relevance of these documents. If the documents are used and referred to, the quality of RAs and control measures will be improved.

Supervisory personnel must be trained on soft skills like communication and team leadership so that they can effectively tap into the knowledge and wisdom of the workers under their charge to effectively prevent accidents. As can be seen from the first two cases, workers on the ground can foresee possible accidents, and failure to listen to workers' concerns can result in accidents. Supervisory personnel should address all safety concerns through the RA process. When a worker identifies a hazard, the work team should discuss the potential consequences and their likelihood in detail and systematically so that suitable measures can be implemented to prevent accidents and ill health. Such onsite and quick RAs force the supervisor or foreman to take a systematic and detailed look at the task instead of addressing the hazards intuitively, during which they may be prone to cognitive biasness such as over-confidence.

Case C showed that although workers do play an important role in preventing accidents and ill health, management and supervisors must still implement a wide range of measures to manage these unsafe behaviours. The WSHA, aligned with WSH management principles, clearly stipulated that workers are expected to cooperate with their employer or principal and the occupier. Thus, when workers continue to disregard WSH measures and the management's best intention to prevent accidents and ill health, the workers must be removed from the workplace or only allowed to work on low risk tasks. Such drastic actions must be taken to prevent routine violations from developing in the workplace.

The range of possible hazards and accidents is very wide. Thus, when evaluating and learning from accident cases, it is important to consider the underlying factors which are more fundamental and applicable to most workplaces. However, most investigations do not have the resources to probe the management system, culture and leadership issues in sufficient depth. As can be seen from the three case studies, only subjective inferences can be made. There is a need for more thorough investigations and the findings must be shared widely to facilitate learning and prevention of similar incidents.

Review Questions

1. Based on the case studies, discuss the importance of frontline WSH leadership.
2. What are the possible ways to influence supervisors' safety behaviour?
3. In the three case studies, do you think the individual workers should be taken to task?
4. How would you apply the concepts in the earlier chapters to change the culture of the organisations in the case studies?

Reference

Reason, J. (1997). *Managing the Risks of Organizational Accidents*. Aldershot: Ashgate.

Index

www.ingramcontent.com/pod-product-compliance
Lightning Source LLC
Chambersburg PA
CBHW070503160726
48003CB00004B/1393